D0930681

CREATIVE AND COLOURFUL CONTAINERS

CREATIVE AND COLOURFUL CONTAINERS

GRAHAM STRONG AND CLAIRE PHOENIX

CAVENDISH BOOKS,
VANCOUVER

Published in 1998 by Cavendish Books Inc.

ISBN 0 929050 93 2

Colour separation by Bright Arts (HK) Limited
Printed by Toppan Ltd
Commissioning Editor Helen Griffin
Typeset and edited by Joanna Chisholm
Designed by Maggie Aldred
Publisher Anne Wilson

CAVENDISH BOOKS,
VANCOUVER

Contents

INTRODUCTION

Fashions and trends in gardening come and go with great regularity, but rarely has there been such sustained interest as exists in container gardening – whether in hanging baskets, windowboxes, pots or tubs. That the popularity of container gardening grows from strength to strength is not really surprising, when you consider its advantages. You have complete control over the end-product, as you can ensure optimum growing conditions. New labour-saving techniques, an ever-increasing range of tempting seasonal and hardy plants and a remarkable array of beautifully made containers have done much to fire gardeners' imaginations. Add to this the fact that there is no strenuous digging and unlikely to be any weeding, and you have what we believe to be one of the most enjoyable, relaxing and creative hobbies.

In *Creative & Colourful Containers* we show you how to achieve a display that you will be proud of and that will considerably enhance your house and garden. We have grown and photographed over 60 recipe ideas in tried and tested combinations, picking out the very best to delight the eye, the nose and even the tastebuds. There are schemes to grow from seed or young plants, tray or pot bedding for a quicker picture and sizeable, off-the-shelf shrubs and herbaceous plants to lend almost instant maturity to your containers. And it doesn't all happen just in the summer. There are ideas to breathe life into your display from autumn right through to the last days of spring.

Look out for some stunning colour schemes, wildflower creations, fun projects for children, containers to scent the garden, and edible tubs and baskets. Feel free to add your own unique ingredients. We know you'll be surprised – and delighted – with the results.

Dwarf botanical tulips like these 'Johann Strauss' look captivating in a shallow bowl teased open by sunlight and underplanted with violets.

THINGS TO CONSIDER

It is comforting to know that you can work miracles with plants in containers, adding instant colour with a tub under a shady tree or with a hanging basket high up on a stark brick wall where a ground-cover plant or climber might have taken years to establish. A permanent framework of planting might look good in your tubs and windowboxes, supplemented perhaps by pockets of bright bedding and a pair of smart clipped box or bay trees to flank the front door.

When planning your outdoor furnishings, you need to ask yourself two pertinent questions: what do I want, and what have I got? It is also vital to consider practical as well as aesthetic factors: a stone trough is likely to be far too heavy for a tiny balcony; a dazzling display of gazanias will fail in the shade; and don't expect your hanging baskets to look a picture if you are away in high summer and don't make alternative watering plans.

Inevitably, expense comes into the planning, too. Box spires for example are costly yet long-lasting; bedding plants are cheap but need replacing two or three times a year.

Think also about co-ordinating your display by repeating a colour scheme or range of plants in a windowbox, tub or hanging basket. Does it need to look good throughout the year, for only the summer or even perhaps for a special anniversary or wedding weekend. If you're beginning to savour the prospect of creating your own mobile garden, here now is a picture gallery of ideas to inspire you.

CONTAINER STYLES

You can grow almost any plant in a container and the right plant can be persuaded to grow in almost anything that holds soil, though the end result may sometimes be twee (stone boots) or even vulgar (toilets).

I also draw the line at white-emulsioned car tyres as suitable vehicles to carry a lovingly tended mobile garden. Strangely enough I find some bright coloured plastic pots as appealing as mellow, weathered stone and terracotta. Plastic urns and cauldrons can look acceptable if you are working on a shoestring budget or are worried about theft from a front garden. In the end, your choice of container comes down to what sort of house you live in, where it will be sited and to individual likes and dislikes.

However, matters of good husbandry are not open to such debate, for as well as aesthetic appeal it is important to consider how practical a container is for growing plants in. This may sound a bit illogical if it is made specifically for this purpose, but shallow bowls and urns will be suitable only for a limited range of permanent plants. Most terracotta wall pots contain such a small volume of soil that they need regular watering (perhaps twice a day) and carefully chosen plants to do well.

Some containers cost a king's ransom but increase in value, while others can be bought for next to nothing or even begged for free. Here are some novel ideas to get you thinking.

• You can draw a chair up to this outdoor fireside when summer evenings begin to get chilly. French marigolds, coleus and lantana will raise the temperature when grown in a fire basket. Line it with black plastic and put fire-blackened logs down the front to add authenticity.

• Old coppers, which are much sought-after as planters, in time develop a wonderful, blue patina. Such a big container needs bold treatment. Here, some blowsy, yellow tulips tower above blue and yellow polyanthus and trailing ivy to create a visual feast.

• Ali Baba jars add mystery – inviting visitors to peep inside – but if you do decide to add greenery choose trailers to cover the top and tumble down the sides. Plant a pot that fits snugly on the rim. Here, coils of dried seaweed have been added to the succulent sedum.

• Troughs as big as this home-made one are roomy enough for a brilliant bedding display and a climber or two at the back as a permanent fixture. The framework is made from 7.5 cm (3 in) x 5 cm (2 in) timber clad with half-round logs lined with heavy-gauge polythene.

• Painting and stencilling are popular ways to personalise a wooden container, but when it's a well-weathered, oak barrel like this one, it's a shame to cover all that natural ageing. Here, fir cones – secured to the rope with screw eyes – have been hung along the front.

• Black plastic cauldrons are roomy enough to grow three or four plants and light enough to lift up for a table display. This combination of wild pansy with *Nemesia denticulata* 'Confetti' and grey-and-white-splashed foliage is a real winner.

● Dark green or brown plastic urns always seem preferable to the ubiquitous, glaring white ones, and the wider and deeper they are the better your plant display will be. For a summer-long show, try a mixture of heathers, coleus and ajuga to colonise the rim.

● These imported, shallow terracotta planters are like giant pudding bowls, superb for packing in summer bedding plants, but likely to get lost at ankle height. Raise them up using an urn pedestal or a terracotta pot and they'll look as good as this one does.

● Buff slabs are the most popular choice for patios so it makes sense to pick out troughs and pots made from similarly coloured reconstituted stone or concrete. Various finishes are available such as Cotswold (like this) and a paler Bath stone.

● Leaky, pensioned-off, galvanised or enamel buckets make ideal homes for plants with a wandering root system like mint or these Chinese lanterns. However, don't be tempted to paint the surface – the weathered look is all the rage!

• Terracotta wall pots can be used in the garden in the same way as you hang up pictures to decorate the lounge, though with such limited room for soil they dry out very quickly and it may be worth considering an automatic watering system to keep them happy.

• Shallow terracotta pans are popular with bonsai growers, but they also look great pushed end to end and planted with the same bedding plant in a different colour. These snapdragons are edged with ajugas, and the pans have been stood in front of lavender.

• Small woodland plants like cyclamen will thrive on an old tree stump or hollow trunk if you keep it constantly moist. Be careful, however, that the rotting wood does not introduce the dreaded, black, thread-like strands of honey fungus into your garden soil.

• Chimney pots are perhaps the tallest containers we can plant and as such are invaluable to give instant height amongst clusters of pots or at the edge of a border, as here. A pair, planted like this with trailing nasturtiums and petunias, are perfect flanking a door.

Unusual Windowboxes

From bicycle baskets to trugs, watering cans to brightly coloured plastic pots there are more ways to arrange plants than in a traditional windowbox.

From pottery to plastic, stone to wrought iron, there are more types of windowbox than you can imagine. A stroll around a nearby garden centre or along the streets of your home town can give you ideas of the containers you would – or perhaps would not – like to use. You may come across a wonderful old lead windowbox or an excellent plastic fake. Each type of windowbox has its advantages and drawbacks. Stone is beautiful but extremely heavy, plastic is lightweight but not so attractive and can age badly, cracking and yellowing. However, when planted, plastic can be indiscernible from, say, terracotta and costs far less. Terracotta looks good and, being heavier than plastic, is more stable; it generally lasts much longer than plastic, too. It is also porous, however, which means that the compost can dry out in hot weather, and is not always frost-proof.

Consider the plants you are planning to use. You may need to adjust your plans so that box and plants complement each other. There's no need always to opt for a purpose-made box or basket. Be a little inventive and use an old hamper or assorted tins and you could create an individual plant holder that will be greatly admired.

• Make the most of old galvanised watering cans by filling them with plants and arranging them along a sill. You will need to drill holes for drainage. Choose trailing plants such as these 'Surfinia' petunias and verbena, although the taller clary stand up effectively.

• An old bicycle basket makes a good windowbox, enhanced by the mass of pelargoniums. Line it with slit plastic, compost and slow-release fertiliser. Choose a sunny spot for pelargoniums and water regularly in hot spells. Hang by the leather straps on screwed-in wire.

• A rustic log provides a harmonious container for year-round windowbox arrangements. Here, chrysanthemums, marguerites and the warm variegated foliage of a coleus provide late summer and early autumn colour.

• A simple line-up of coloured pots in nature's clashing colours provides a refreshing contrast to more organised arrangements. From left to right, heliotrope, viola, oxalis, busy lizzies, pelargonium, marguerites, marigolds and plectranthus all create a pleasing array.

• Fill an old trug with a mass of soft colour and sit it on a sill. Here fuchsias, busy lizzies (*Impatiens walleriana*), petunias and floss flowers mingle delicately with silver-leaved *Plectostachys serphyllifolia*.
Tip: Use plastic liner to keep the wood from rotting.

• Plant up an old beer crate by lining it first with slit plastic and then add crocks and compost. This version had a slatted base but you may need to drill drainage holes. It is deep enough for a scaevola, scabious, petunias and marigolds to bloom happily.

HANGING BASKETS WITH A DIFFERENCE

From old kettles to colanders, unusual baskets are tucked away in corners just waiting to be filled with flowers. Seek a change from the traditional wire basket.

The earliest type of hanging basket was made of simple wires strung together in a lattice pattern. Now there are a great many different types, from plastic bowls on chains, with a reservoir to aid watering, to ornate 'bird cages', which make an attractive feature even when the basket is empty.

The most popular option is a plastic-coated wire basket, which does not rust – unlike earlier versions – and has plenty of spaces for planting from below. There are plastic versions of this type with vertical fins, which make inserting plants even easier. There are also purpose-made hanging balls, which break in half for planting up and then clip together again afterwards; in the past, two standard baskets had to be wired to form a sphere.

Wicker baskets can look extremely effective but are best lined with plastic to prevent excessive leaching of water and compost. Unusual containers, such as colanders and kettles, make good hanging baskets and are worth looking out for.

• Original cast-iron cauldrons, such as this, are widely copied in lightweight plastic. If you have an original, make sure your fixings are secure as they are weighty. A blend of oxalis, chrysanthemum and pelargonium here creates a pleasing container.

• Make the most of an unusual basket – this one is made of green wire and bamboo – by planting harmonising colours. The simple theme of white *Dianthus* and variegated ivy is particularly striking and does not detract from the container.

• Purpose-made wall pots are available in glazed and unglazed pottery – they make good holiday souvenirs. This terracotta version has a moulding of grapes and looks just right when filled with oxalis, verbena and petunias, as here.

• A metal colander makes a pretty and unusual hanging basket. With drainage holes ready made, you simply need to hang it, using strong wire or hook-on chains. Trailing petunias and *Convolvulus sabatius* tone in picturesquely with this blue-and-white colander.

• A three-tier, wire vegetable holder may be out of vogue in the kitchen but makes an excellent hanging basket. Here three apricot portulaca, with their fat, waxy leaves, make an eyecatching layered arrangement that thrives best in full sun.

• Hunt out an old kettle or oversize teapot for a more unusual hanging basket. Trailing plants such as *Lysimachia lyssii*, petunia and verbena combine with more upright marigolds and blue felicia for a colourful arrangement that makes passersby smile.

• Windowboxes are cheering on balconies where they please both those inside and out. Hanging baskets can look particularly attractive when colour-balanced with boxes below. Here fuchsias, ivy-leaved pelargoniums, lobelia and busy lizzies create a striking show.

• A typical town house. Here grey brick and stonework have been lifted with bright windowboxes of red pelargoniums (*P. hortorum* 'Red Elite'). A basket of trailing begonias, fuchsias and pelargoniums links with a border of floss flowers and more pelargoniums.

• Dahlias at the doorstep, hanging balls of trailing pelargoniums (*P.* 'Mini Cascade') and cascading fuchsias combine with a myriad plants in the border to make this row of terraced cottages one of the prettiest exteriors seen – or rather hidden.

• A simple, pale grey frontage has been given a lift with a mass of trailing, pink pelargoniums. To ensure such a wonderful display, you need to begin early in the season – giving the plants a good start. A large, deep windowbox to retain moisture will also help.

CREATING AN IMPRESSION

Against any building backdrop, whether brick, stone or render, windowboxes and hanging baskets are a celebration of the season.

Wherever you live, a hanging basket or windowbox will make a lasting first impression on visitors.

This is your opportunity to create a highly welcoming windowbox or basket that says how much you care about your home. Decorating need not stop at immaculate paintwork – baskets and windowboxes will soften the exterior of your home and give it an inviting and perfectly finished appearance. A carefully tended windowbox or basket will bring pleasure to you and passersby for a whole season and, with carefully chosen planting, can act as a link between your home and garden.

Before selecting any plants, consider the exterior of your home. Will your basket or windowbox be mounted against brick, stone or render? Do you wish the colours of your plants to contrast or blend? Could you plan a scheme to tone in with the paintwork or other house features or do you intend the plants to stand out in a bold display of colour?

Take a look at your neighbours' gardens and houses. Do you wish to create a complete contrast or do you want to plant boxes and hanging baskets to tone in with the 'look' of your street. Don't just think of spring and summer. Lovely baskets and boxes can be planted for year-round enjoyment.

• In a riot of colour, this bungalow has been given a vibrant facelift with balls of busy lizzies and marigolds, baskets of pelargonium and lobelia and a contrasting border of floss flowers and marigolds that is vivid enough to stop you in your tracks.

• This unusual cast-iron windowbox of pink, trailing pelargoniums would be appreciated from both inside and out. On each side of the garage entrance, baskets containing four varieties of fuchsia, pelargoniums and lobelia brighten what could be an otherwise dull area.

• Two brimming baskets and a stunning bank of summer flowers lift these garages out of the ordinary. Petunias, pendulous fuchsias and marigolds subtly enhance the pink of the rendered walls while softening the harder edges of the building.

• Strong pink trailing verbena and pelargonium with stunning red busy lizzies and begonias are mixed with shades of lobelia (*L. erinus* 'Rosamund' and *L.e.* 'Crystal Palace') to create a wonderful balance of colour against this blue door.

• Glorious baskets of pelargoniums – both upright and trailing – petunias and lobelia. Prettily variegated ground ivy (*Glechoma hederacea* 'Variegata') and the light foliage of helichrysum echo the pale green door. Remember that flowers in full bloom can be bulky.

• A porch doorway is framed and softened with three baskets of pelargoniums, petunias, silvery helichrysum and lobelia. The colour scheme is followed through at ground level with wooden tubs filled with flowers in colours that complement the basket plantings.

MAKING AN ENTRANCE

Create a welcoming first impression with brimming baskets hanging by your door.

If the windows are the eyes of your home, then your doorway is its mouth and voice. The secret of these planting schemes is that they have all been carefully chosen to flatter the doorway and its surrounds. From a double garage to a disused cellar, all doorways benefit from complementary colour.

One of the joys of small-scale gardening is that, even when using similar plants, each basket and box is totally individual. The popular summer bedding plants – pelargoniums, petunias, fuchsias, busy lizzies and lobelia – are each time arranged in differing proportions, using an extensive range of varieties. Employing blue-and-pink-toned plants to echo a door colour makes an attractive arrangement that is easy on the eye.

Contrasting foliage plants add further interest, with the result that each basket is unique and can be planned to reflect the character of your specific home.

Obviously, it is important to think ahead a little before fixing hanging baskets by a doorway. A basket that has just been planted takes up no more room than its own diameter. When in full bloom, however, the total area it occupies is considerably more. For this reason, do not set the baskets too close to the doorway, otherwise the petals and foliage will be damaged by people brushing past.

• An unused doorway looks inviting with this collection of pots and a bright hanging basket of marigolds, verbena and lobelia beneath the arch of a leafy wisteria. Make sure that you do not allow foliage and plants to obscure a background feature too much.

• A wooden door is given a fresh country feel when flanked by two colourful baskets. Lobelia, fuchsias, pelargonium and gold bidens combine with the garden to create a welcoming entrance. Without the baskets, this doorway might look rather severe.

WINDOWS

Plant your windowbox to make the most of your window or to change its character completely.

Some windows will benefit more from hanging baskets and windowboxes than others. If your house is close to the road, has little or no front garden and many passersby, a basket or windowbox can have a noticeable effect and will be much admired.

Even if your windowbox is for your eyes alone, you will receive months of pleasure for relatively little outlay. It is worth planning the position of your hooks and boxes before you start. Take note of successful arrangements in your area and you will discover which sizes and combinations work best. You may find that windowboxes would work better above, rather than below, your window, provided they are easy to water.

One long windowbox is less likely to dry out than two smaller ones. You may wish to have a number of windowboxes or baskets arranged vertically rather than horizontally. Symmetrical baskets often work well, although it is rare to see such symmetry in nature! A number of small hanging baskets will need a great deal of care as these tend to dry out quickly; fewer, larger baskets may be a better choice and watering will prove less of a chore.

Just as in art, everyone's taste in flowers differs. You may wish for single-colour planting, mixed colours or something completely different such as vegetables. Consider the next section, on season, size, soil and situation, before you make your final decision.

• Pretty baskets on each side of a bow window differ in size and planting schemes. Nasturtiums and pelargoniums on the right link with petunias in the barrel below and contrast with fuchsias, petunias and trailing variegated ground ivy in the basket opposite.

• Strongly shaped planting suits an individual window. The spiky cabbage palm has been surrounded with box, cyclamen, gold-tinged euonymus, ivy and silver *Senecio cineraria*, all of which will survive winter in a sheltered spot and continue to provide interesting foliage.

• A bay window with deep sills has the advantage of three separate positions for windowboxes. You could opt for a co-ordinated scheme or, as here, aim for three completely different plantings of ivy, red pelargoniums and marigolds with lobelia and helichrysum.

• White render makes an excellent backdrop for a traditional mix of pelargoniums, begonia and lobelia. The unusually positioned boxes above the window link well with the ground-level planting and the pristine newly painted white lattice above and below.

• Sometimes delicate planting schemes can be the most effective, especially when they complement their background – in this case an old brick and half-timbered building. Here a mix of the palest pink petunias and mauve and white lobelia works well.

• A cascading windowbox links ground floor and basement windows. Three pelargoniums – a double bright red (*P.* 'Gustav Emich'), a bicoloured pink and a small white pelargonium – sit above purple verbena and trailing deep pink 'Surfinia' petunias.

THINGS TO THINK ABOUT

Before you rush out and buy a basket or box and some colourful plants to fill it, there are a few things worth remembering. They could be called the four Ss – season, size, soil and site – all factors worth bearing in mind whether you are an expert or novice at container gardening.

● Be aware of the size of your house in proportion to the size of basket. Too small a basket, such as this one of pink and white diascia, looks lost, whereas the basket of the correct size and planting will enhance your home and appear to be in balance.

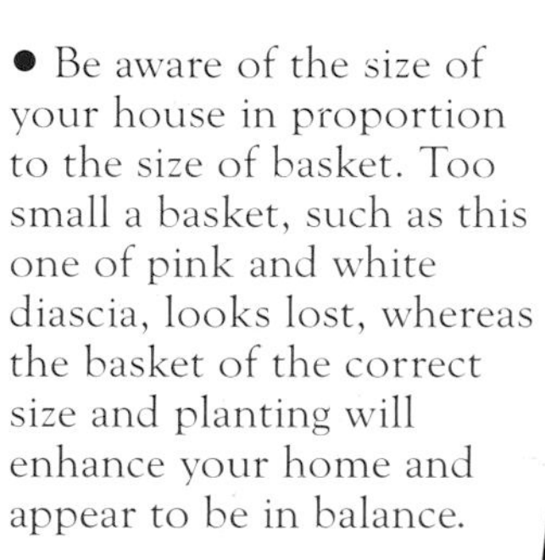

Season

It used to be assumed that hanging baskets and windowboxes were just for the warm summer months but now you will see colourful pansy balls throughout autumn and early spring and wonderful winter windowboxes that are especially welcome when gardens are bare.

Evergreen foliage can be attractive in itself – with golden tones and berried plants adding extra brilliance. Later in the year, some plants may be discarded and flowering plants added.

Size

Do bear in mind the size of your window when buying or making a windowbox. Those that are too small can look rather mean and out of place. The larger and deeper your box, the easier it will be to maintain healthy plants. A larger box can hold more soil and therefore retains more moisture, thus avoiding one of the major problems – that of drying out.

Think about buying a basket a size larger than you had first planned – 30 cm (12 in) baskets are really suitable only for very small arrangements. A 35 cm (14 in) basket is a good general standard, while a 40 cm (16 in) basket will give you a full and professional-looking display. The larger baskets are very heavy when the compost is wet, so make your fixings strong and secure.

Soil

Some plants need a particular soil type to survive. Many heathers, for example, like an acid soil and will quickly die in limey soil. Most plants are quite happy in a multi-purpose potting compost but it is worth checking.

Some plants, such as alpines, need well-drained soil, which can be achieved by mixing in horticultural grit. If you are planting them in a windowbox, place a layer of broken crocks or

• A windowbox of hot summer colours that thrives in a sunny spot: marguerites, tagetes, petunias, helichrysum, lobelia, pelargonium and felicia all love the sun, whereas the busy lizzies would do equally well in semi-shade.

pebbles over the holes in the base of the box to ensure adequate drainage.

Most plants dislike sitting in water and if you are worried about overwatering it is worth layering charcoal in the base layer of soil to prevent mould. You will find further information about compost types on page 188.

Site

It is easy to assume that hanging baskets and windowboxes are best situated in a sunny spot. However many plants, such as busy lizzies, fuchsias and pansies, will flower happily in the shade and numerous evergreens will give a verdant display in shady spots all year round.

Success often depends on the position of the basket or box. Check whether plants prefer a sunny, shady or semi-shady position before selecting them for your container display.

• These pink and white ornamental cabbages and heathers create a living collage that will last through the long winter months. Although many heathers need ericaceous (acidic) potting compost, these heathers are lime tolerant – and cabbages also like lime.

• Here the pale green of spotted laurel and the darker green of dwarf conifers form the basis of an attractive winter scheme – cyclamen will survive among the warmth of city buildings, as will *Senecio cineraria* and the trailing variegated ivy.

POSITIONING YOUR POTS

Now we've looked at container styles, colour schemes and ideas for sun and shade, you will probably be well on the way to composing your own display. You have a mental picture. Now what about the picture frame?

Putting your containers into an appropriate setting can double their impact. Like choosing the décor for a room, your pots are part of the garden furnishings and it's important that they don't strike a discordant note. Simple, hand-thrown, weathered terracotta pots demand a suitably modest setting. Ornate urns and classically decorated pots lend themselves to more formal treatment spaced perhaps equidistantly along a balustrade or straight pathway, or limited to just one imposing planter set up on a pedestal. It's also important to consider scale as well as style. Small pots can be swallowed up and lost in the open garden whereas, when grouped together, they not only catch the eye but are also easier to maintain.

Positioned on stark expanses of paving or brickwork, it's hardly surprising that many containers fail to realise their potential. Far better to site a random assortment of containers as individuals enveloped in planting, raised up just enough to show off the top half of the container or in a sea of gravel where carpeting alpines and herbs can spread freely around the base. You do however need easy access for what is often daily maintenance.

• These pots of French marigolds stand out from the crowd and frustrate the attentions of slugs and snails by resting on pedestals made from clay chimney liners and upturned pots. Even a terracotta rhubarb forcer will provide a good perch in summer.

• As an alternative to a conventional windowbox, a parade of small pots lined up on a windowsill and planted with pansies or busy lizzies looks charming from inside and out. Aim for uniformity and select pots of the same size and shape.

• As an inexpensive and eye-catching way to flatter a large container, eight bricks can be laid like the spokes of a wheel on a bed of gravel. In time, the pansies will seed into the gravel. If you're impatient, why not plant some pansies amongst the fragrant creeping thyme.

• Stone troughs like this are usually planted with alpines and stood on four bricks on the patio, but this one has been ingeniously built into a stone, retaining wall. The profusion of lady's mantle and the greenery 'anchors' it to the spot.

• Against colour-washed walls it is always tempting to develop a Mediterranean theme. Props like this tall honeycomb jar add to the atmosphere, as do lavender, scented-leaved lemon verbena, fleshy agave, yellow marguerites and the red bottlebrush flowers.

• A roomy, wooden wheelbarrow makes a picturesque home for a cluster of pots filled with early spring flowers and buds. Being mobile, it can be sited around a doorway or even under the house windows for armchair viewing. The pots are insulated with straw.

• A bold group of three large pots of similar shape but contrasting sizes makes more impact than 30 tiny ones scattered at random over the patio. These blue-glazed Chinese pots were planted with a permanent framework of evergreens supplemented by seasonal bedding.

• A flight of steps make a wonderful, tiered stage setting for pots, provided they are built wide enough to allow the edges to be decorated. Machine-rounded logs (for risers) and granite setts (for treads) are ideal materials for making curved steps.

• A bare wall can be almost instantly turned into a feature by the clever use of wall pots. Group them together for maximum impact, avoid straight lines and don't forget to include a comfy seat where you can sit back amid your handiwork.

• An outsized container like this has enough presence to stand in splendid isolation as a centrepiece in a courtyard or at the end of a vista. The planting should be equally bold and stick exclusively to one species, like these superb tobacco plants.

• This trio of pots planted with succulents and succulent-leaved alpines sits on a plinth of Yorkstone and pebble paving retained by a brick edge. It forms a small terraced area overlooked from the house and successfully highlights this collection of special pots.

• This handsome terracotta urn is one of a pair guarding a flight of steps – a classic arrangement that rarely fails to please. For a more dramatic contrast with the billowing, shrubby potentilla, use purple-leaved cabbage palms in the urns.

• Occasionally a cluster of pots needs a little extra something to give it a lift. This stylish terracotta pineapple provides just such an irresistible focal point amongst cottage-garden favourites like lady's mantle, pinks, lavender, golden feverfew and houseleeks.

• An edging of pebbles gives this collection of busy lizzies, coleus, echeverias and pelargoniums an informal, country appeal. You could also run the pebbles as a mulch around each pot to hold them steady and visually anchor them to the ground.

WHERE TO HANG A BASKET

The most obvious place for a basket or pair of baskets is by a doorway – and with good reason. It can be seen whenever you enter the house and will be enjoyed by neighbours and casual passersby.

However there are many other areas that benefit greatly from a colourful display of baskets. Fences and walls, sheds and garages all become attractive features when festooned with flowers and foliage.

You can make your own holders for baskets – such as the post with brackets and the trellis shown here. Or simply use standard brackets well fixed into the wall.

Consider where your basket will be most appreciated. A herb basket by the back door will be in easy reach and will bring the scent of aromatic herbs into your house.

Try to plant according to the position of your basket. Choose sun-lovers for a sunny site and plants tolerant of shade for cooler spots.

Wherever your basket is to be hung, check that it is not going to hit anyone on the head. Baskets can be extremely heavy when full and well watered.

Do ensure that your basket is not going to be a nuisance to pedestrians – if it is going to hang over a pavement, set the fixings higher and use a rise and fall system for watering.

• Two rows of baskets along a fence cleverly disguise what could be a boring area of the garden. *Lobelia erinus* in shades of purple, lilac, white and the less common, deep carmine 'Rosamund' vie for space with marigolds, pansies, alyssum and striped petunias.

• Create a striking basket of pastel colour using one simple variety. Here busy lizzies (*Impatiens walleriana*) hung above a doorway combine tones of pink, white and lilac planted all round the basket to give a exuberant ball effect.

• Transform a dull brick wall with a trellis of hanging baskets. Here brilliant pelargoniums (*P.* 'Paul Crampel' and 'Queen of the Whites') are intermingled with petunias and marigolds underslung with red and blue lobelia interspersed with variegated ivies.

• A fence post or purpose-made bracket holder creates a stunning area of the garden when holding two brimming hanging baskets. Striped petunias, white alyssum, pelargoniums and marigolds make a show when mingled with mixed shades of lobelia.

• Fuchsias, busy lizzies and lobelia all tolerate a shady spot – such as under this tree. The colours in this arrangement, especially the lilac, purple and masses of white lobelia, combine to create a cool and delicate spherical basket.

• Make the entrance to your house welcoming with brimming baskets of colour hung beside the door. Delicate pink tobacco plants, lobelia and grey-leaved helichrysum tone with the stronger pinks of 'Surfinia' petunias, busy lizzies and fuchsia in the adjacent basket.

• Symmetrical baskets of trailing ivy-leaved pelargoniums (*P.* 'Galilee'). To create a spherical effect, two baskets have been planted separately and then wired together. This shows how baskets can contrast effectively with your wall colour, rather than toning in.

• A striking purple and stained-glass door is set off to perfection with a basket of petunias, lobelia and variegated ground ivy, in harmonious shades of pink, green and lilac, backed by rambling clematis bearing purple flowers that match the shade of the door.

• It's tempting to echo the colours of your garden in your windowboxes but here a cool planting scheme of green foliage would simply fade into obscurity. Instead, a bright arrangement of marigolds, busy lizzies and petunias provides a refreshing contrast.

• Five pots of vibrant dwarf gerbera have extra impact when used en masse between tomato-red window shutters. Do not be afraid to employ bold blocks of colour. Try veering away from pastel colours and you are likely to be surprised and delighted with the results.

COLOUR BALANCE

From co-ordination to creative contrast, the plants you use in your boxes and baskets should be chosen carefully. Plan them to link with your house or garden and the result will be tastefully artistic.

The colours you choose for your planting schemes are your own personal area of design, even on such a small scale, so it is worth trying to imagine the effects you hope to achieve rather than just picking up a random assortment of plants at your garden centre and hoping that they will all work together 'somehow'.

First, look at the exterior of your home. Are there colours you can pick up on, e.g. in the door or window frames? Are there plants in the bed below the window, which you would like to echo in your windowbox or hanging basket?

This is not to say that a mixed selection of plants cannot look particularly effective. Indeed, most baskets that you will see contain a mixture of colours. By studying other baskets, you will see that some of the most effective are those that have been planted to co-ordinate with, or make a contrast to, the house, another windowbox or each other.

Plan ahead. If you think your paintwork or render needs freshening up, do it before you hang baskets or fit boxes. Of course, if you are thinking of repainting, you can always make your new paintwork match your windowboxes, baskets or other containers.

• A cascading basket of pendulous fuchsias and variegated ground ivy echoes and emphasises the pink wall. Here verbenas, busy lizzies and variegated pinks on the windowsills link with the healthy hydrangea below to stunning effect.

• Numerous shades of yellow and orange plants work well together and contrast attractively here with the green foliage and paintwork. Marigolds, pansies and celosia co-ordinate beautifully with variegated ivies, helichrysum, coneflowers and sunflowers below.

• An array of these feather duster plants would brighten up any corner. Sometimes nicknamed Prince of Wales feathers, this dwarf strain of Plumosa group *Celosia* comes in red, pink, yellow and salmon and looks amusingly pert with its crested blooms.

• A striking combination of marigolds, busy lizzies, marguerites, pelargoniums and petunias. Take a closer look to spot the blues of felicia and lobelia, the white marguerites and deep red lobelia – the details that give the arrangement depth.

HOT COLOURS BRIGHT IDEAS

Make the most of nature's brilliance with a basket or windowbox in vibrant colours that cheer and inspire. Choose from a profusion of begonias to a cascade of mixed summer colour.

Many of the plants that enjoy the heat of summer come in a range of hot and sultry colours. From begonias and the fiery feather dusters of celosia to luminous marigolds and bright pelargoniums, their brilliance proclaims that summer has arrived at last.

Many daisy-like plants, including osteospermums, gazanias, zinnias and helichrysum, open up in brilliant sunshine. Others that particularly enjoy direct light and warmth are petunias, pinks, miniature roses and verbena. Direct sunlight can, of course, be terribly fierce and drying, so it is vital that you select appropriate plants if wanting to site a container in such conditions. Some plants will need watering more than once a day and perhaps moving to a shady spot for a few hours to avoid the worst of the sun's rays.

Succulent plants, such as portulaca and cacti, are especially designed to survive well in hot conditions, storing water in their fleshy leaves.

Most plants sited in very hot places will need plenty of water. Avoid getting water on leaves or petals as the tiny drops of water act like magnifying glasses in strong sunlight and will cause the leaves or petals to burn.

• For the first bright, heartwarming colours of the year, you can't fail to please with ranunculus – wonderful peony-like plants from the buttercup family. Available from early spring in rich single colours and pink-edged white, they will flower for several weeks.

• In return for rich moist potting compost, begonias will reward you with vibrant colour. For windowboxes choose the smaller-bloomed low-growing varieties, such as Multiflora begonias; do not plant out until the frosts are over. Here, a terracotta pot adds to the warm effect.

• The apricot-yellow of pendulous begonias is given extra warmth by the proximity of double pink petunias. Purple brachycome and pale pink lobelia combine with pink and white busy lizzies to make this a harmoniously colourful basket.

• Luminous 'Non Stop' begonias, French marigolds (*Tagetes patula* 'Goldfinch') and large clusters of purple lobelia (*L. erinus* 'Crystal Palace') combine to provide an eyecatching windowbox that will remain a profusion of colour throughout summer.

• Create a miniature garden of soft colours in a windowbox. A clipped box sits beside a trellis 'fence'. Miniature roses grow among pelargoniums and parsley. Trailing bugle (*Ajuga*) and variegated lemon balm (*Melissa officinalis*) soften the line of the box.

• Delicate shades of an unusual lavender (*Lavandula lanata*) combine with the cool greens and creams of variegated sage, ivy and a small-leaved helichrysum (*H. petiolare* 'Roundabout') to create a windowbox that will give off a delicious scent when brushed past.

COOL CREATIONS

Aim for subtle, calming colours – greens, yellows, pinks and whites – for baskets and windowboxes that suggest peace and tranquillity. Blend them together for a harmonious scheme, choosing a range of delicate shades from the same colour palette.

In contrast to many of the fiery schemes of high summer, it can be particularly refreshing to cast your eye over something a little more subtle. There are many wonderful foliage plants that tend to be ignored or used solely as a backdrop. However, when used together, these can look sophisticated and quite different in a cool quiet way.

Many herbs have an interesting leaf shape, which makes them useful both artistically and for culinary purposes. Flowering thyme, the soft 'rabbit ears' of sage and the feathery foliage of fennel give textured appeal as well as the added bonus of a Mediterranean fragrance wafting through the air.

Most of the silver-leaved plants, including *Senecio cineraria* or dusty miller (*Lychnis coronaria*), need more sun to keep their colour than might be imagined, but are particularly rewarding.

Some miniature roses can be obtained in very sweet, soft colours, as can a whole range of wildflowers.

• The arching stems of this attractive grey-leaved double marguerite reach skywards and create an attractively feathered hanging basket. An upright fuchsia (*F.* 'Sunray') and grey-leaved pelargonium keep the scheme within similar cool tones.

• In contrast to many brightly coloured windowboxes, the soft yellows and whites of violas, marguerites and busy lizzies and the subtle blue scaevola offer a refreshingly natural look. These plants would enjoy a position in semi-shade.

• For a simple but effective, winter windowbox for a very sheltered spot, purple cineraria (*Pericallis* x *hybrida*) surrounded by evergreen foliage (spotted laurel and a star-shaped ivy) make a striking combination. The bold cineraria flowers for many winter months.

• For an interesting foliage basket that also produces pale blue flowers from mid-spring to early summer, bugle (*Ajuga reptans* 'Burgundy Glow') is a good choice. With its wine-red leaves shaded with white and pink, this evergreen will grow happily in sun or deep shade.

• Many tall plants also come in dwarf varieties, which are suitable for windowboxes. These blue dwarf delphiniums look particularly attractive against a lemon-washed wall with the green foliage echoed in the box. They flower from late spring to early summer.

• Miniature 'Tête-à-Tête' daffodils herald the arrival of spring, their strong, fresh yellow heads nodding in the breeze against verdant, standard and attractively variegated ivies. Plant the ivies in autumn with daffodil bulbs below and winter pansies above.

• A mass of trailing, pink pelagoniums (*P.* 'Mini Cascade') provides welcome colour on a smart city street. The black door and shutters contrast sharply with the soft pinks in the hanging basket. Pelargoniums are ideal for baskets as they tolerate a little neglect.

• An ever-popular mix of pendulous fuchsia, begonia and trailing pelargonium will win favour through being prolific and striking, although not without some effort. Fuchsias are particularly demanding of water. This basket is suitable for a semi-shady position.

SINGLE-COLOUR PLANTING

For baskets and windowboxes with the greatest impact, it is hard to better the effect of single-colour planting. Hanging balls or overflowing windowboxes of one flower or colour type never fail to catch the eye.

It is extremely easy to push a trolley around a nursery or garden centre picking up trays of plants that catch your eye, perhaps even buying two or three new varieties if you are feeling adventurous. The results, however, may not always be as impressive as you might hope. Take a look in other trolleys and you will see that some people are planning an all-blue scheme or one of white and green only. Even before planting up, you can see how well colours of a similar hue work together, while a mixture can look garish and jumbled.

Take a tip from the experts and plan your scheme before you venture out. Make notes from year to year of successful and rewarding schemes and keep the plant tags to remind you. From sunny yellow daffodils with simple green foliage to an all-white daisy, or all-blue lobelia, ball, there are a myriad ways to use a group of plants within one area of the colour palette. The results can be most effective and well worth the extra time taken in planning your plantings.

• For a strikingly pretty ball of daisies, follow the planting instructions for the pansy ball on page 160. If plug plants are not available simply make larger holes for more mature plants. The marguerites will fill out and, with regular deadheading, will bloom all summer.

• A wicker hamper makes an attractive windowbox when lined with slit plastic to prevent compost from leaking out. Here a mass of orange chrysanthemums provides a block of warm colour that will flower from late summer to early autumn in a sunny spot.

• Green and white schemes look effective – clean and modern, romantically soft and pretty – especially when many plants are used. Here New Guinea busy lizzies, pinks (*Dianthus* 'Snowflake'), ivy and *Ficus pumila* 'Variegata' are artistically combined.

• You can grow herbs on a sunny windowsill but this basket would need to spend some time outside too. Sorrel, parsley, mint, chives and thyme would cater for many of your cooking requirements and the mingled, trailing foliage looks particularly attractive.

• This relatively new fuchsia variety, *F.* 'Autumnale', has wonderful leaf colouring and creates enough foliage interest to make a striking hanging basket on its own. However, it also produces bright pink flowers in summer and autumn and enjoys a shady position.

• As an alternative to ivy, plectranthus is rapidly gaining in popularity for windowboxes and hanging baskets. Its attractive variegated leaf and trailing habit make it an ideal backdrop to many planting schemes. Plectranthus also looks dramatic when used alone.

SHADES OF GREEN

Rethink your attitude towards foliage schemes – the subtleties of leaf structure can be more fascinating than the most colourful planting arrangement.

When planning a hanging basket or windowbox, many of us instinctively opt for the latest trend in colours – recently, blues, pinks and purples have had a revival, whereas ten years ago oranges and yellows were in vogue. Looking at some of the windowboxes used by commercial properties such as hotels, shops and offices, however, proves that there are valuable lessons to be learnt. Here a backdrop of evergreens, such as conifers and ivies, is used all year round. These evergreens are then interplanted with polyanthus and pansies in spring, pelargoniums and begonias in summer, chrysanthemums in autumn and cineraria in winter – all prolific-flowering plants of their season, which look good against a green backdrop.

It is unfair to place foliage plants in the sole role of backdrop to the flowering planting schemes, however, as many non-flowering plants are very rewarding in their own right.

From the luxuriant colours of coleus to the deep burgundy-brown of Japanese maple (*Acer palmatum*) there are innumerable foliage plants that go extremely well together or will create an unusual hanging basket when used entirely alone.

• A multitude of plants has been chosen here, particularly for their unusual foliage. From bloody dock to nasturtiums, zonal pelargoniums to hosta, pick-a-back plant to periwinkle, the different leaf structures look wonderful when grouped in a windowbox.

• For an attractive trug of green foliage plants that is also useful, keep herbs on a sill near the kitchen door. Flowering thyme, variegated sage, parsley and tarragon all jostle happily for space and will give an aromatic bonus too when carried on the warm summer air.

STRIKING COLOUR SCHEMES

There are plenty of ideas here to tempt you, some taking their prompt from the colour of the container or even an individual, two-tone flower.

If you've never thought of colour theming your pots, you may be surprised to know that the less you put in, the more you get out of your containers – at least as far as the number of colours is concerned – though you do need a lot of self-control to stick to just one. If you're a little unsure as to what goes with what, why not limit your ideas to two or three tried and tested colour blends: blue and yellow, pink and blue, orange and blue, and so on.

In spring you can usually get away with more outrageous colour combinations, because there is so much greenery in the garden to soften the impact. In summer there is an explosion of possibilities and the chances are you'll come up with something unique and, if you like it, who cares what others think – though I've yet to see an orange-and-magenta success story. Even in wintertime there are good pairings to be had. For example, in a container you can bring foliage and flower into such close proximity that they will often have twice the impact of the same plants in a border.

You can also select colours to suit your mood. Cool blues and white are just the thing to refresh you in the heat of the day. Alternatively, use yellow flowers to create artificial sunshine on dull days and add reds and oranges to really turn up the heat. It's great fun. Why not try it?

• Sometimes the colouring on the faces of pansies is so special that it's worth giving them a setting all of their own. These *Viola* x *wittrockiana* 'Melody Sunrise' were planted in a nest of chocolate-coloured heuchera and ajuga leaves to highlight the blooms.

• If you pick out the richest colours from the summer bedding sales benches and gather them together like this you will have a really sumptuous show. Here we have *Salvia coccinea* 'Lady in Red', red pelargoniums, yellow African marigolds and red and blue petunias.

• A coloured pot like this blue-glazed bowl is a good starting point for a special theme. The selected flowers range through palest blue (*Pulmonaria* 'Highdown') to the rich, velvety primrose (*Primula* 'Midnight') and the exquisite, silky *Crocus vernus* 'Purpureus Grandiflorus'.

• Unusual summer annuals like these golden daisies (*Ursinia anthemoides*) or *Salvia coccinea* 'Lady in Red' are easily raised from seed. This is well worth the effort as they can rarely be bought. Mix in some silver and purple leaves for a striking colour scheme.

• You can create some real drama in late spring by sticking to a colour scheme using yellow and maroon. The pansy was the starting point, then the yellow early double tulips and double primrose were added and, for foliage, red houseleek, grey sedum and bronze fennel.

• It becomes progressively more difficult to make a splash towards the end of summer, but asters can always be relied upon to create colourful interest. The colour pink dominates this planter with ornamental cabbages, mixed dwarf asters and *Sedum ewersii*.

• A trough can become a feast of colour by alternating purple and pink winter heathers backed up by pink and blue hyacinths. To ensure they peak together, pot up your hyacinth bulbs in autumn and force them under glass or buy pot-grown bulbs once in bloom.

• Double daisies and violets are diminutive charmers for late spring and their colours never seem to clash. This blend of pink, purple and yellow is particularly pleasing – and a great recipe for the front of a windowbox or trough.

• These spectacular South African heathers are a great way to pep up flagging container displays before the winter heathers come into full bloom. They fit in well with blue-leaved conifers, but don't expect them to survive severe frosts unless very well protected.

• With their often outrageous colours, tulips can provide some of the best opportunities to conjure up a bit of plant magic. *Narcissus* 'Tricolet' here tones in beautifully with these *Tulipa* 'Garden Answers', orange wallflowers and trailing *Euphorbia myrsinites*.

• Imported glazed pots are usually frostproof and beautifully coloured. This mid-blue bowl is crammed full of spring violets and the cup-shaped daisies of *Anemone blanda*. A touch of yellow adds contrast without losing the harmony between plants and container.

• Pastel pinks and blues are soft, restful and easy on the eye. Here blue *Campanula poscharskyana* trails into a pink-flowered marguerite, over the glaucous leaves of dicentra planted below and through yellow-foliaged physocarpus growing in an adjoining pot.

• The ragged leaves of ornamental kale 'Red Peacock' have veins and midribs of a lighter purple-pink. Here, white-fringed petunias, whose flowers pick up the kale leaves to perfection, are growing up through its leaves to give a summer-long show.

• Orange flowers can be refreshing and vibrant when combined with purple and have never looked better than in this clever mixture of pansies in a reconstituted stone pot. Don't let the unpopularity of orange as a bedding-plant colour frighten you from using it.

Single plant types or species

Simplify your ideas and you will be surprised by the results. Individual plant schemes have a great deal of poise and horticultural panache.

While you can achieve a planting scheme of great impact by using a number of different plants within one range of colour, the most striking baskets and windowboxes, especially from a distance, are those containing a single species, variety or plant type.

Whether you choose a windowbox of begonias or a basket of busy lizzies, the results are likely to be impressive. Check through our pointers on pages 24 and 25 to ensure that you have taken into account the four Ss – season, size, soil and site. Busy lizzies will thrive in a shady spot, but some plants need more careful siting. Different varieties of pansies bloom happily throughout the season, but not all plants are so content all year round. Some plants, such as the blue hydrangeas shown here, are rarely seen in hanging baskets, perhaps because they require ericaceous (lime-free) potting compost as well as large amounts of water and possibly also because they can grow fairly large, so that most people think of them as a garden shrub rather than a container plant.

The easiest single plant types to grow must be pelargoniums, more varieties of which are being bred all the time. As single-planting schemes, they are hard to better.

• Pansy balls make rewarding hanging baskets, bringing bright colour to cheer the winter months, and then again through spring and summer. Numerous varieties are available such as these cheerful *Viola* x *wittrockiana*, Universal Series. (Follow the instructions on page 160.)

• Bold pelargoniums (*P.* 'Red Elite') brighten the grey brick of a typical town house. Pelargoniums flower best in a sunny spot and are easy to grow as they can survive with less water than many plants, although occasional feeding will boost the display.

• Linking two baskets, one hanging, one used on a windowsill, with a similar colour scheme, looks effective without being ostentatious. The ball of pink tradescantia harmonises well with the chrysanthemums – yet both have individual character.

• Create a stunning, spherical basket using the colour spectrum of one single species. Here busy lizzies (*Impatiens walleriana*) have been used to great effect by combining subtle tones of pink, white and lilac planted through the basket to give all-round growth.

• Enjoy a blue mophead hydrangea at close quarters by siting it in a basket with variegated ivies. You can buy the ericaceous compost which it needs to keep its colour or convert multi-purpose compost by adding a liquid solution, using powder from a garden centre.

• The palette of pink, white and purple shades found in these heathers (*Erica cinerea* 'Moonstone White', *E. vagans* 'Mrs D.F. Maxwell' and *Calluna vulgaris* 'Silver Knight') combines well in a windowsill trug. These heathers require damp ericaceous compost.

HOT SPOTS

Plants grown in containers positioned on paving or beside a house wall are particularly vulnerable in sunny weather.

If long, hot summers and water rationing are here to stay, it therefore makes sense to choose plants that revel in such conditions. Busy lizzies and tobacco plants, for example, are not suitable as they'll need watering twice a day.

In the increasingly Mediterranean climate we are experiencing, many of our gardens are beginning to take on a rather exotic flavour, with pots brimming with daisies and architectural foliage plants. For plants that come from countries like Australia, Mexico and South Africa, it is becoming a real home from home as they have revelled in the hot sunshine.

Several tell-tale signs indicate the sort of plants that are able to withstand exposure to hot sun and periods of drought, although, when restricted in a pot, only a handful will thrive on just natural rainfall. Fat, juicy leaves (as on many succulents) are one such indicator and narrow, spiky foliage (as on yuccas, cabbage palms and grasses) is another. Hairy or grey-felted leaves (as on pelargoniums and lavender) will also reduce water loss.

Drought-busters are especially useful for colonising shallow containers that hold only small quantities of soil or for pots in inaccessible places such as on a high wall or perched on a brick pier. When you do need to water, use one of those slightly curved lances that fit on the end of a hosepipe to give you extra reach into the container.

• Carpeting thyme is well suited to drought conditions and searing heat, and three varieties that contrast in leaf and flower create a classic composition for this shallow, terracotta pan. Finish off the display with a dressing of fine chippings, and site the pan in full sun.

• The houseleeks in this shallow, cast-iron urn will survive on just rain from the heavens, though you will get much juicier rosettes if you water two or three times a week in hot weather. Add plenty of grit to the potting compost to ensure good drainage.

● Plants with a rosette of narrow, sword-like leaves will often tolerate a few days of missed watering. They also lend an exotic atmosphere to the garden, particularly when grouped like this purple cabbage palm, agave succulent and yuccas, creating a feast of foliage.

● Many members of the daisy family are real sunseekers, like these blue kingfisher daisies (*Felicia*) tumbling past a decorated terracotta pot filled with grey-leaved echeveria succulents. They make a lovely combination on top of this retaining wall.

● When you've got a pelargonium as exuberant and distinctive as 'Frank Headley', it's worth giving it a special setting. This wooden planter picks up the sage-green colour in the pelargonium's white-variegated leaf and is one of a pair flanking a matching bench.

● Zinnias are stunning in a pot all to themselves when the sun shines, but they have a reputation for sulking in wet summers. Don't sow the seed too early. Mid May is soon enough, in an unheated greenhouse or indoors on a light windowsill.

Shady corners

Although foliage plants are the most obvious choice for containers set in shade, a host of bedding plants like fuchsias, begonias, busy lizzies and monkey musk also relish a cool spot out of the sun's glare.

A lot of pot-grown, spring flowers such as primroses and anemones are well adapted to life in the shade of hedgerows and woodland. In summer, they just tick over or die down below soil level. Finding summer-flowering plants that tolerate shade is far more likely to be a problem. When robbed of sunlight, plants such as marigolds and geraniums produce more leaf than flower and grow tall and spindly.

It's important to be aware of the difference between shade cast by overhanging tree branches and shade from a wall or fence, where the indirect light levels will be far higher and consequently your choice of shade lovers much enhanced. A pot-grown plant will often grow better beneath a tree canopy or at the base of a wall than one planted directly in the ground, where it has to compete for water and nutrients. Some trees such as yew, however, do cast particularly heavy shade while others such as lime and sycamore may drip sticky honeydew on plants below.

Green and blue leaves generally do better in shade than variegated ones, which may revert to plain green. Some of the largest-leaved plants grow in semi-shade, where an increased surface area is needed to gather all available light, so again look to nature to point the way.

• This superb, purple-leaved Japanese maple is an ornament that any gardener would love to use to give character to a shady spot. In this particular situation, however, the leaves could be spoilt by sticky honeydew dripping from a birch canopy above.

• Fuchsias, like this 'Swingtime', will thrive at the base of a north-facing wall. Take the opportunity to build up a display that will lighten up the area, as well as conceal the bareness of the slabs and walls, by adding trailing lobelia and a yellow-leaved pick-a-back plant.

• You can make a really welcoming frontage to your home, even if it gets no sun, by maintaining a collection of potted evergreens (here *Rhododendron yakushimanum*) and Japanese maples supplemented by lush foliage plants, like hostas, that excel in containers.

• Some foliage plants are quite capable of making their mark without association with other species. The brown spring leaves of *Rodgersia podophylla* look splendid on their own in a wooden tub mulched with cobbles, and will double in size in a month.

• Plants such as fuchsias and busy lizzies are excellent for providing flower colour in shady areas. For an eye-catching, banked display, position them in a cluster of three pots of differing heights, fuchsias to the back and busy lizzies in the front.

• Ferns are amongst the first plants that spring to mind when we think of shade. A collection of three with contrasting leaf shapes makes a fine trio for clay pots, and they will grow contentedly for years as long as they are not allowed to dry out.

Using Spring Bulbs

Like fireworks, bulbs are ready-made surprise packages. You just plant them, stand back and wait for the show. They are tailor-made for pots and with a bit of advance planning you can double your spring colour.

Thumbing through those glossy bulb catalogues in late summer and autumn, or drooling over pre-packed wall displays at your local garden centre or nursery is like being let loose in a sweet shop. The choice is almost overwhelming and all will light up your garden from the first stirring of snowdrops and crocus in winter to the last tulips of spring. For exposed sites, it's best to avoid tall floppy daffodils and tulips, which are easily blown over. Hyacinths too may flop over the edge of the pot as they age, but that is all part of their charm.

The secret when buying bulbs is to aim for continuity so that, like a relay race, several bulbs may come into flower together, overlapping with bedding plants such as primroses and pansies. Then, as they fade, new varieties are ready to carry on the show.

If you're clever, you can manipulate your potted bulb collection to defy the seasons by bringing some into a greenhouse and holding back others in a cool shady spot outside so that for an anniversary or wedding you will have a real flower festival of your very own creation – far more satisfying (and cheaper) than buying them all in flower.

• Winter heathers are perhaps the most valuable of all types, but why be content with a sheet of pink, white or purple when you could have twice the impact by planting dwarf bulbs, like these crocus, in-between their wiry stems to make an embroidered carpet.

• Early double tulips are amongst the best choice for small to medium-sized containers. They are sturdy, compact and available in a myriad of colours. 'Fringed Beauty' is a really hot customer, especially with red and yellow wallflowers and brassy yellow, dwarf doronicums.

• You can sometimes develop a theme based on the colour of your container. This vegetable tureen is ideal to play host to scented *Iris reticulata* and white crocus, though you must water carefully as there is no drain hole, and site it out of rain to avoid waterlogging.

• Pansies really get into their stride in April and May, and De Caen anemones make lovely companions to have alongside. They are vibrant, tough and easily grown. Plant them between the pansies as dry tubers in autumn or in bud in spring from smaller pots.

• Every plant has a perfect partner and colour matching is one of the greatest joys of gardening. Try this blend of white, yellow-eyed primroses with cheeky blue *Anemone blanda* for a cool, sophisticated show. Both can be set out in a border after flowering.

• Scale is an important factor to consider with container gardening. When you plant small bulbs, like these 'Tête-à-Tête' daffodils, surround them with plants of similar stature (here crocus, crocus tulip, mixed primroses and *Veronica peduncularis* 'Georgia Blue').

Foliage finesse

Foliage plants have many moods conjuring up visions of steamy jungles or dry, arid landscapes, and most will thrive in a pot.

After indulging in all that explosive outpouring of summer colour, isn't it nice to gaze out on a cool oasis of luxuriant foliage, although some yellow-leaved evergreens may provide a ray of sunshine throughout the year. Many of the larger-leaved kinds will grow to surprising proportions given a big enough container and, of course, plenty of feeding and watering.

Pot culture also enables us to grow tender exotics like bananas and coleus on the patio in summer and return them to the warmth of a conservatory, greenhouse or well-lit room before the first frosts. Their leaves will continue to give pleasure even after the most prolific flowering plant is spent.

The fun begins when you start to group together your plants so they offer maximum contrast with each other: spiky leaves with broad, rounded shapes; soft, feathery foliage with bold, jagged-edged leaves; yellow with reds, greys with purple. For real drama, site one of the extroverts, like giant rhubarb, in splendid isolation in a half barrel to make an irresistible focal point. In its second year you should be able to rest in its shade, though don't imagine that you have to grow giants to make your mark. There are some exquisite, small-leaved foliage plants amongst ferns and ornamental grasses and don't forget dual-purpose plants such as astilbes and bergenias that have handsome flowers and leaves.

• The explosion of luscious leaves in spring and early summer can be enjoyed at close quarters in pots stood on gravel. Here, deadnettles and epimedium vie for attention with the almost black-leaved *Anthriscus sylvestris* 'Ravenswing' and dwarf ornamental rhubarb.

• Even on a dull day, a yellow-leaved evergreen like this *Choisya ternata* 'Sundance' will inject a bit of artificial sunshine and really cheer you up. A member of the orange family, it has scented leaves and flowers, and needs warmth and shelter in cold areas in winter.

• For a classic foliage threesome that will succeed even in a shallow tomato tray like this, try *Hosta* 'June', *Carex oshimensis* 'Evergold' and *Houttuynia cordata* 'Chamaeleon'. Slow down the mint-like spread of the houttuynia by keeping it in its pot.

• If you fancy a show of coloured foliage that will really leap out at you, mix together the most extrovert, yellow-variegated leaves you can find, combining three or even four shades. Perhaps the most useful plants are those that also have bold flowers.

• Ornamental cabbage and kale represent some of the most exciting newcomers for seasonal bedding and are particularly bold in late summer and autumn, when cooler temperatures encourage the rich pink and purple leaves to intensify in colour.

• A collection of leafy herbs can be sited in a pot or two alongside your favourite garden seat. Run your fingers through them to release the delightful fragrance from pineapple sage, chamomile, apple mint, lad's love and lemon-scented pelargonium. Absolute bliss!

Country style

In a garden you can deliberately go out of your way to deceive, because even if you live in a town or city it's still possible to manufacture a small piece of idyllic countryside.

Mixed hedges of thorn and hazel, wild flowers creeping in at the boundaries and apple trees weighed down with fruit go a long way towards achieving the country look. Containers, too, can play a big part – and who wouldn't want to own a small slice of the countryside, even if it is only encapsulated in a few pots? The look is achieved in two ways.

Firstly, your choice of plants is important. Cram your pots full to their brim with annuals and perennials that will shoot skywards and spill over the rim. Spiky New Zealand flax and cabbage palms will never work; they are far too exotic. Some would also rule out variegated leaves or bright yellow and sombre purple foliage and, indeed, in tree form they can jar against a backcloth of rich greens. Highly bred plants such as petunias and red salvias can also look too man-made for an authentic country look.

Secondly, look out for nicely weathered pots without elaborate decoration – simple, hand-thrown terracotta, shallow baskets and sun-bleached oak barrels. Stone troughs are also good, though the genuine article is now very expensive. There are, however, some very convincing concrete reproductions available, which are distressed to look old. Stand your pots on salvaged brick or stone flags or on increasingly sophisticated, simulated versions of these.

• This patio is a mirror image of the cottage wall behind and a great backdrop for a container show. Long-tom terracotta pots were chosen to house yellow lysimachia, sweet William and marguerites but beware on windy sites as they may blow over and crack.

• In this subtle country scheme, a pot and flaking door set the theme for a subdued mix of love-lies-bleeding and tobacco plants intricately woven through with floss flowers and brachycome. Yellow creeping Jenny softens the edge of this composition.

• This brick wall makes a splendid location for a collection of cottage-garden favourites like sunflowers, hollyhocks, pansies and herbs. Don't make it all too clean and tidy. A bit of decay, a few yellow leaves and plants encroaching on each other enhance the look.

• Violas are highly collectable, diminutive charmers, and these rustic shelves are just the place to show them off and bring the scented varieties closer to the nose. Stack the clay pots in decreasing sizes and station your biggest plants at the base so the shelves don't look top heavy.

• Sun-bleached oak and rusting bands of iron add character to this wallflower display. The white brickwork and weatherboard behind throw each flower into sharp focus, and the colours and perfume of the wallflowers seem to instil the very essence of a country garden.

• Wicker log baskets filled with lavender and flowering perennials are a lovely way to welcome visitors to your front door. You don't have to plant them permanently. These lady's mantle, Canterbury bells and sweet Williams are still in their original plastic pots.

PERMANENT PLANTING

Some plants are so well adapted to life in a pot that they can be seen as virtually permanent fixtures and if well maintained will add character and maturity amongst more fleeting bedding plants.

Growing plants in pots is addictive and you'll soon have an ever-expanding collection in all shapes and sizes. Rather than continually changing each one with summer bedding plants followed by spring bedding and bulbs, it's a good idea to plant up some with a permanent framework of evergreens, such as conifers, rhododendrons, camellias and heathers, to clothe the patio in winter. Herbaceous plants like hostas, dicentras and hellebores with good leaves and flowers will also earn their keep from year to year amongst the more fleeting annuals.

Permanent plants don't have to be grown as isolated individuals, each in their own container. Larger pots, troughs and tubs can have two or three permanent, hardy plants around which the bedding and bulbs are set out. Alternatively, once a camellia or rhododendron has finished flowering it can be moved away from the main display area and brought out again the following year to repeat the performance.

Container-grown, permanent plants are also at a premium where surfaces such as tarmac make it impossible to plant a climber in any other way, and a pair of potted and clipped box or bay trees flanking a door is still a design classic.

• There are some exquisite varieties of dicentra that have delicate, often grey-green leaves and pendant, locket-shaped flowers that last for weeks in spring and summer. Here 'Pearl Drops' and 'Stuart Boothman' have been in pots for two years in a semi-shaded area.

• During the dead of winter, there is sometimes little but evergreens to fall back on. However, conifers of contrasting colour and form when combined with dwarf hebe, as here, can look really cheery. With a powdering of snow, you could almost be in the Alps.

• These hardy slipper flowers bloom after the two spring-flowering evergreens (rhododendron and pieris) have finished, while variegated hostas and purple-leaved bugle carry the interest forward right through summer and contrast well with the shrubs' new growth.

• Formal, square casks or Versailles tubs never look better than when planted with a clipped box spire like this or with a pyramidal conifer like *Thuja occidentalis*. The tubs can be painted to match any patio trellis or even house colours if used as a pair to flank a doorway.

• A stone trough looks better when permanently planted with alpines or slow-growing, truly dwarf conifers (like these) – or perhaps a combination of the two – than with a riot of nasturtiums or petunias that would quickly hide its weathered surface.

• Some 'permanent' plants are less long-lived than others. After a couple of years these violets will need replacing with new ones, but the beautiful rock rose, pink heucherella and fleshy sedum should last twice as long before they need to be transferred to a border.

THEMES AND SCHEMES

Have you ever wondered how nurserymen manage to produce such perfect uniform plants brimming with vitality? Their secret is blueprint growing, following tried-and-tested methods of cultivation. We have applied the same principles to container gardening and devised 60 recipe ideas for all sites and seasons. Many are effortlessly achieved – important if for example you want to nurture a child's interest (pages 110–13) and it's also hard to fail with cosmos and petunias (pages 134–5). Follow the step-by-step instructions and you will be pleasantly surprised.

Container gardening is not just about growing perfect specimens though. Marrying plants to their container, developing a clever theme or restricting the rainbow of colours can really set you apart from the crowds. You can be cooling off amongst luxuriant foliage in the heat of summer, distilling the very essence of the cottage garden into a clay pot or windowbox, or dining out on home-grown salads without moving from the sun lounger. In winter, imagine drinking in spicy perfumes at nose level just outside the back door.

Finally, remember that you can literally rise above your existing wet sticky border or limey soil by growing plants that have previously always been off-limits. Those delicate spring alpines that demand good drainage will thrive in a trough or basket, and delectable lime-haters like the pieris on this page can be readily grown in lime-free potting compost. What are you waiting for? The sky's the limit!

Making a Windowbox

Begin by measuring and marking the position of the angle brackets, as shown. With the bradawl make holes, where marked, for each set of four screws. Insert the screws – an electric screwdriver makes the job much quicker. Ensure that the last piece you add is a long side and screw the brackets to this before fixing it to the box as it can be fiddly screwing both angles of the last brackets directly into the box.

Lime is dangerous to work with as it can cause caustic burns unless you wear thick gloves. If you do choose to use powdered lime, you will need to add water until you have a runny liquid. A similar effect can be achieved using white emulsion.

You may need to water the paint down until it is the consistency of single cream. Apply to all areas of the box, except the base. When dry, use the wire brush to remove paint until you are happy with the result. A certain amount of woodgrain should show through. Treat softwood with wood preservative to prevent bowing and splitting.

YOU WILL NEED
8 stainless steel angle brackets,
5 cm (2 in) x 5 cm (2 in)
3 pieces timber,
15 cm (6 in) x 75 cm (30 in)
2 pieces timber, 15 cm (6 in) x 15 cm (6 in)
32 x 1 cm (3/8 in) flat-head screws
tape measure · pencil
screwdriver · bradawl

1 Follow the instructions in the main text for assembling the windowbox. Drill three holes at regular intervals in the base for drainage. Increase the number of drainage holes if you are making a larger box.

2 If you are staining the box, rub in woodstain using a soft cloth. We chose a soft yellow to enhance our bronze stars and silver moon stencils but almost any shade of brown or green would suit a floral arrangement.

3 Old, roughened timber is often best for showing off a limed effect. You may need to rough up the grain with a wire brush, deepening the grooves along the straight of the grain. Some woodgrain should show through.

4 Stencil the limed trough using your chosen design, or plant it in its natural state without any further ornamentation. A coat of clear matt varnish will ensure that your limed effect and stencil do not wash off.

Stencilling

Plastic stencils are widely available in craft shops or you can make your own using a scalpel and stencil card. Quick-drying stencil paints are available but thick household emulsions can work just as well on wooden windowboxes as long as a final coat of clear matt varnish is put on to seal in your design. If you use exterior paints, this will not be necessary.

You may have leftover household paints which you can use or you might be able to obtain small sample pots that fit your intended scheme. Small pots of enamel paints in primary colours, and metallics intended for toy painting are also suitable.

Many gardeners have artistic leanings and you may be inspired to paint your own container, such as the watering can shown here. First paint your container in a base colour and allow it to dry completely. Then, after sketching your idea on paper, transfer it to the metal medium using suitable paints. For beginners, an abstract leaf shape would be easier than the scene shown here of swans on a lake. If this is the first time you have painted pots and containers, do not be too ambitious with your first attempts.

YOU WILL NEED

1 litre (1¾ pt) base colour exterior paint – matt, silk or gloss (for box) · a stencil masking tape · chinagraph pencil enamel or gloss paints · stencil or thick-headed brushes · fine brush for detailed work

1 Using plastic or terracotta pots, fix on your stencil with masking tape and apply the paint using a stencil brush. If stencilling a wooden windowbox, paint it in your base colour; allow it to dry thoroughly overnight.

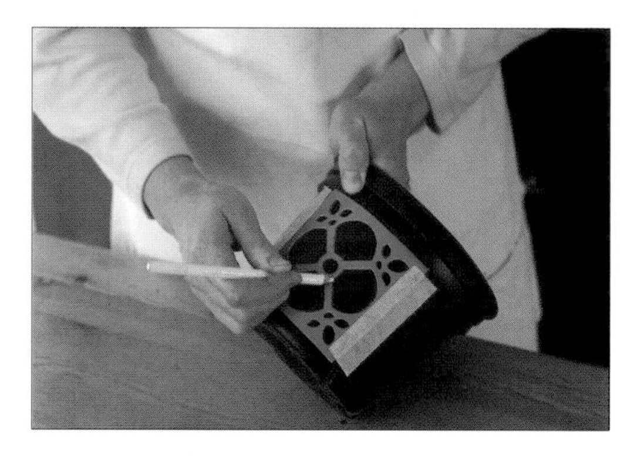

2 If the paint tends to seep beneath the stencil on a plastic pot, it may be easier to outline the design using a chinagraph pencil. Then remove the stencil and infill with enamel paints using the smaller paintbrush.

3 You may wish to use just part of a larger stencil to create a co-ordinating theme. Bold blocks of primary colours work well. Fine detail will only be seen close up so decide in advance where to put the pot.

GOOD ADVICE

- Choose stencils that give bright blocks of colour. These will look very effective from a distance.
- If paint runs beneath stencils on a plastic pot, outline the design in chinagraph pencil and then paint this.
- Do not be too ambitious with your first attempts unless you are sure of your artistic skills.
- If painting a windowbox, apply a painted base coat of one colour and allow this to dry overnight. Otherwise the stencil may smudge.

Hyacinth Parade

Compared to the poise and elegance of a dwarf daffodil or tiny wild cyclamen, a hyacinth can come across as a bit of a clumsy heavyweight. It has a stiff rigidity that would strike a discordant note naturalised in grass for example. I don't care a bit, because in many of the best garden plantings an understanding of a plant's personality is the key to its success. A hyacinth excels in a formal setting so I lined up these 'China Pink' at the back of a stoneware trough. With crocus and 'Wanda' hybrid primroses they looked rich and regal and their intoxicating perfume wafted on the slightest breeze. Although not quite as free flowering as the florists' varieties, 'Wanda' hybrid primroses are hardier and make better garden plants. Their leaves are often tinged with purple.

YOU WILL NEED

a reconstituted-stone trough, 43 cm (17 in) x 20 cm (8 in), finished in Cotswold colour
6 pink hyacinths (*Hyacinthus* 'China Pink')
20 crocus (*C. vernus* 'Purpureus Grandiflorus') · 5 mixed hybrid primroses (*Primula* 'Wanda')

PERIOD OF INTEREST
March and April

LOCATION
In sun or shade near an open window or doorway.

FOR CONTINUITY
Plant out the crocus and hyacinths in the garden and replace them with pansies and forget-me-nots.

Plant up the hyacinths and crocus in autumn and transfer them to your trough in spring. Alternatively you can buy pot-grown hyacinths and crocus in bloom in spring. I forced on my pink hyacinths under glass so they were in bloom with the crocus. Fill the trough to one-third with compost. Arrange the hyacinths at the back. Then plant the crocus and primroses. Top up with compost and water in.

GOOD ADVICE

- Think carefully about where you position your trough and get someone to help lift it in place before you fill it.
- Make way for a summer display by planting out the primroses amongst snake's head fritillaries and cowslips in a semi-shaded spot.

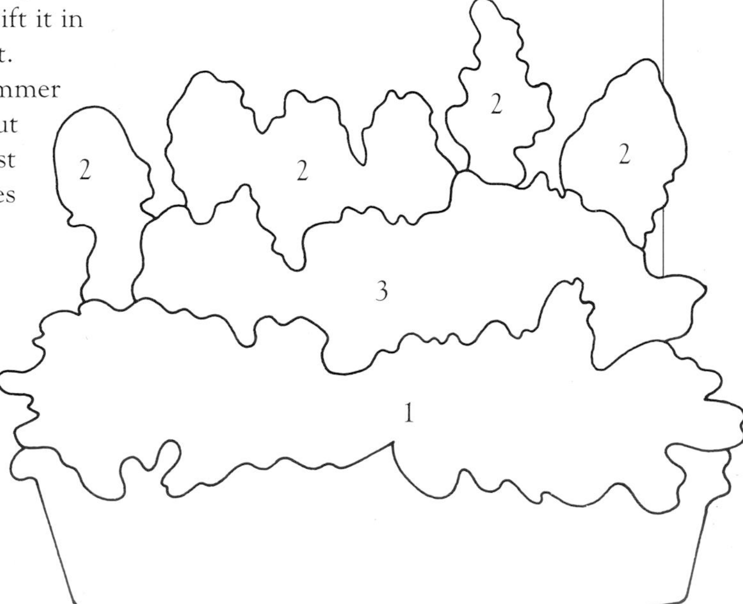

1 *hybrid primroses*
2 *hyacinths*
3 *crocus*

Double Delight

I have always retained a child-like fascination for members of the daisy family. They have an innocence and simplicity about them even when the petals are increased a hundred fold to make a fully double bloom like these spring bedding daisies. Too lavish a container would intimidate them so I chose a weathered garden riddle to show them off. With the ever-increasing use of soil-less composts, however, these sieves are more likely to be seen as ornaments than as an essential piece of gardening kit. They are still made today, or you can beg an old one. Deep-rooted plants like wallflowers would fail to thrive in such a shallow receptacle but fibrous-rooted daisies love it – and you don't even need to buy them as pot-grown plants. If you sow seed in summer and bed them out in a garden border, they will transplant successfully into a container the following spring.

YOU WILL NEED

a garden riddle or punctured, old, enamel washing-up bowl · plastic liner · 10 mixed double daisies (*Bellis perennis*)

PERIOD OF INTEREST
March to May

LOCATION
In sun or shade, set the riddle on bricks or granite setts, in front of a clump of bluebells or a haze of forget-me-nots.

FOR CONTINUITY
Let some seedheads self-sow around the garden and even in the lawn.

1 Slit open an old compost bag to line the riddle and add a shallow layer of potting compost. With a trowel, lift your daisies from the border with a good rootball.

2 Before you replant them, clean up each plant by pulling off dead and yellowing leaves or dead flower stems. Pack them in the riddle tightly and fill in with more compost. Then give them a good soaking to settle them.

GOOD ADVICE

- For a finishing touch, a ring of shells makes a nice nursery-rhyme addition to hide the soil around the edge.
- Instead of raising new double daisies each year, you could transplant them from the riddle into a piece of spare ground after flowering, having split them up. Spray with an insecticide if greenfly appear on the leaves. In autumn transplant them again into borders or pots.
- Alternatively, in summer lay down the double daisies where you want them to seed, cover the roots with soil and young seedlings should appear in a few weeks' time.
- Treat forget-me-nots in the same way as double daisies for a never-ending supply, though the resulting stock may be inferior to the originals.

Spring Medley

I don't mind admitting that I'm a pansy addict. I never tire of their cheeky, impish faces. With other familiar spring personalities like forget-me-nots, grape hyacinths and double daisies they exude the sort of charm not always found in more highly bred plants. These pansies were grown from seed sown in July and overwintered in a cold frame. The grape hyacinths were potted in autumn, but there is no reason why you can't buy them in spring, pot grown and coming into bloom, though it's not quite as satisfying. The proportions look better if you group your spring medley in a shallow container rather than a deep barrel or terracotta pot. A woven basket gives a lovely informal touch to a display that meshes together to form a seductive tapestry of colour.

YOU WILL NEED

a sturdy wicker basket, 40 cm (16 in) in diameter and 20 cm (8 in) deep · plastic liner · 3 double daisies (*Bellis perennis*) 6 pansies (*Viola* x *wittrockiana* 'Gold Princess' and *V.* x *w.* 'Lilac Cap') 12 grape hyacinths (*Muscari*) · 4 dwarf forget-me-nots (*Myosotis*)

PERIOD OF INTEREST
April and May

LOCATION
In sun, on a log pile, balcony or patio table or in a porch or conservatory.

FOR CONTINUITY
Frame the pansies with 5 Junior petunias (*Petunia*, Fantasy Series, Milliflora Group).

1 Line the basket with black plastic, then pierce drainage holes in the liner. Add a layer of potting compost before laying out the plants.

2 Spend some time getting the plants into a pleasing arrangement, daisies to the front, pansies and grape hyacinths in the middle, forget-me-nots mainly at the back.

3 Fill in with more potting compost, taking care not to soil the blooms, then water the plants well. Gently dress the arrangement, lifting up the flowers and draping some grape hyacinths over the edge of the basket.

GOOD ADVICE

- Deadhead, water and feed regularly so the arrangement lasts all summer.
- Transplant the daisies and grape hyacinths to a border after flowering.

1 *double daisies*
2 *pansies 'Gold Princess'*
3 *pansies 'Lilac Cap'*
4 *grape hyacinths*
5 *dwarf forget-me-nots*

Basket Weave

Hay baskets have all the advantages of hanging baskets, such as instant colour where you want it, but are less prone to drying out and give a greater volume of planting. To avoid a hay basket lying empty throughout winter, with its metalwork exposed like a skeletal rib cage, strip out the remnants of your summer bedding in autumn, top up with potting compost and plant the daffodil bulbs. Alternatively for a more instant effect, wait until late February when pot-grown bulbs and spring bedding are available in flower and plant in one smooth operation. I wound in yellow weeping willow stems between the framework to help retain the moss and pick up the yellow theme above.

YOU WILL NEED

a cast-iron or plastic-coated-steel hay basket, 90 cm (36 in) x 28 cm (11 in) plastic liner · moss · prunings from weeping willow (*Salix babylonica*) or yellow-stemmed dogwood (*Cornus stolonifera* 'Flaviramea') · 9 pink and yellow polyanthus (*Primula*) · 20 daffodils (*Narcissus* 'Tête-à-Tête')

PERIOD OF INTEREST
March and April

LOCATION
Shelter is more important than sun as daffodils and polyanthus tolerate shade.

FOR CONTINUITY
Add winter heathers (*Erica*), which will flower before and after the daffodils. Replace the daffodils with buttercups (*Ranunculus asiaticus* 'Accolade').

1 Cut to size your prunings from the weeping willow or dogwood and weave them into the hay basket. Then line the back of the basket with black plastic to keep the compost away from any brickwork.

2 Fill in from behind with tightly packed sphagnum moss – or you could even take some from your own lawn, though don't be surprised if a few weeds appear in the basket later on.

3 Add some potting compost and lay out the polyanthus and daffodils while still in their pots until you have a pleasing arrangement. Tap off the pots, insert the plants into the compost and water well.

GOOD ADVICE

- Push some small-leaved, variegated ivies (*Hedera*) through the moss and willow-stem layer if you want to fill out the front of the basket. These can be retained for summer.
- Plant the daffodils in a border after flowering.

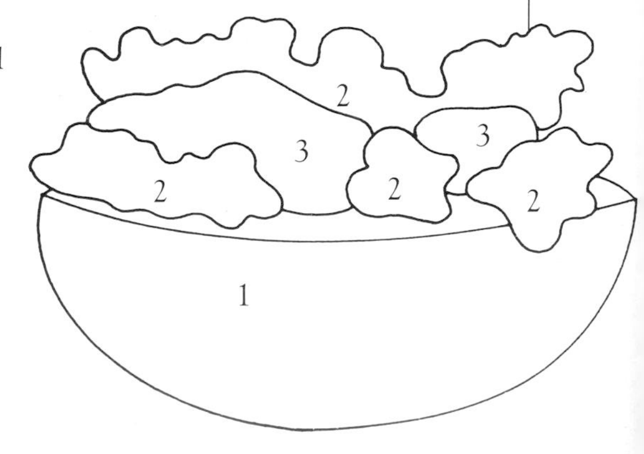

1 *weeping willow stems*
2 *polyanthus*
3 *daffodils*

Mellow Yellow

Overdosing on undiluted yellow flowers – such as forsythia in spring – can make your eyes water. However, as is often the case when colour scheming, a combination of yellow flowers and more soothing, yellow foliage can give a very pleasing display. Add a touch of purple and lime-green in the form of the purple-leaved wood spurge and you have a creation that generates its own sunlight with a very long season of interest. Spurges are among the most useful plants for pots, because they look good for months on end. *Euonymus fortunei* 'Emerald 'n' Gold' also earns its keep throughout the year, although it is at its most colourful as the fresh, new growth flushes in late spring.

YOU WILL NEED

a salt-glazed pot, 40 cm (16 in) in diameter
1 or 2 *Euonymus fortunei* 'Emerald 'n' Gold'
1 purple-leaved wood spurge (*Euphorbia amygdaloides* 'Purpurea') · 1 golden-leaved feverfew (*Tanecetum parthenium*)
2 or 3 black and yellow pansies (*Viola* x *wittrockiana*) · 2 *Helichrysum petiolare* 'Limelight'

PERIOD OF INTEREST
April to June

LOCATION
In sun or shade, though some sunshine will bring out the best leaf colours.

FOR CONTINUITY
Add dwarf tulips amongst the euonymus for April interest and African or French, yellow or vanilla marigolds as a replacement for the spurge in late May.

1 Assemble your chosen plants, including one good-sized or two smaller euonymus plants, and water them all well. Then fill the pot with potting compost to within 5–7.5 cm (2–3 in) of the rim. Plant the centre of the pot with the euonymus.

2 Add the wood spurge. Carefully tuck in the feverfew and pansies amongst the euonymus. Top up the compost and give it a good soaking. Wait until early May before adding the helichrysum.

GOOD ADVICE

- You may find feverfew seedlings in your own or a friend's garden.
- To avoid powdery mildew, spray the wood spurge with a fungicide every 10–14 days.
- Transfer the wood spurge to a border in late May.

1 *euonymus*
2 *wood spurge*
3 *feverfew*
4 *pansies*
5 *helichrysum*

Cool Blue

A special container calls for a special planting theme and I was rather taken with this lion's head trough. It was deep enough to take a specimen-sized pieris, which I had been growing in a pot for two years, with room around the front for more seasonal plants. I gathered up some crocus, hyacinths and double primroses, and soon the blue-and-white theme had become inescapable. Such a colour scheme always looks cool and sophisticated and is quite a contrast to all those unplanned explosions of mixed spring colours you can get if you don't buy your plants in selected shades. As well as the plant range seen here, you can get your blues and whites from *Anemone blanda*, pansies, violets and grape hyacinths, so there are plenty of variations on this theme.

You will need

a reconstituted-stone or plastic trough, 55 cm (22 in) x 40 cm (16 in) · ericaceous potting compost · 2 double blue primroses (*Primula* 'Blue Sapphire') · 1 *Pieris japonica* 'Cupido' · 5 blue and 5 white hyacinths 10 white crocus

PERIOD OF INTEREST
March to early May

LOCATION
In sun or shade, on paving near a window and door or by a garden path where the hyacinths' scent can be enjoyed.

FOR CONTINUITY
Add dwarf *Iris reticulata*, which will flower before the crocus.

1 You can plant up the trough in autumn, buy plants in flower in spring, or copy me and pot up your bulbs in autumn and transfer them to the trough in April. Arrange the potted plants into a pleasing display.

2 Clean up the primroses by picking off any yellow or decaying leaves and spent flowers. Also remove any slugs!

3 Add a thick layer of potting compost to the trough and plant the pieris at the back. Then add the hyacinths and lastly the crocus and primroses. Fill in any gaps with compost and water in.

Good advice

- Use a lime-free, ericaceous potting compost to prevent the pieris turning yellow.
- After flowering, return the pieris to its original pot. Plant the other plants in the border.

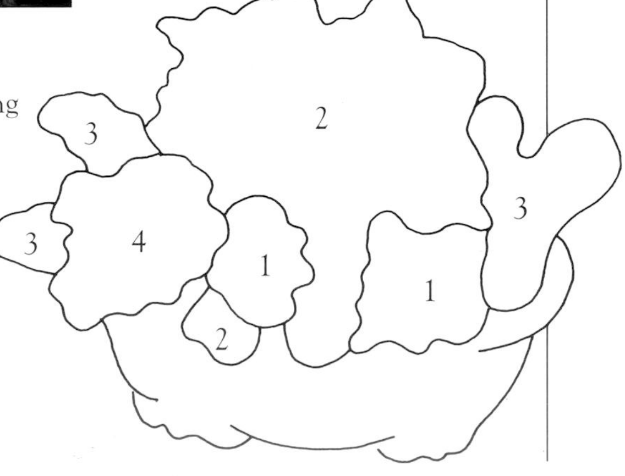

1 *double primroses*
2 *pieris*
3 *hyacinths*
4 *crocus*

Easter Windowbox

The best-known, spring-flowering bulbs – daffodils, tulips and crocus – have earned their popularity by being hardy enough to withstand the cold winds and rain of early spring and by looking bright and colourful at a time of year when our gardens are only just coming into bud. However, many lesser-known bulbs are also worth trying, from the bell-like blue scillas to anemones.

As many spring bulbs flower a few weeks after each other, many gardeners suggest surface planting with pansies or primulas to give a constant display through which the bulbs can grow.

Our colourful windowbox is quite simple to plant, using trailing ivies, box and rosemary as a foliage base, with blue grape hyacinths underplanted in autumn and primula and polyanthus in yellow, pink and purple added later.

YOU WILL NEED

a plastic trough, 75 cm (30 in) x 15 cm (6 in) · 25–30 grape hyacinths (*Muscari*) · 1 box (*Buxus*) 1 rosemary (*Rosmarinus*) 2 trailing ivies (*Hedera*) 2 purple polyanthus · 3 primulas

PERIOD OF INTEREST
December to April

LOCATION
In full sun or partial shade.

FOR CONTINUITY
Add in Persian buttercups (*Ranunculus asiaticus*) as the grape hyacinths and primula fade.

1 Cover the drainage holes in the base with crocks. Fill the box to just below the rim with potting compost and plant the grape hyacinth bulbs in clumps 2.5 cm (1 in) below the surface in mid- to late autumn.

2 Plant the box and rosemary to the rear and the ivies over the front and sides. Infill with polyanthus and primulas, leaving space for the grape hyacinths to push through. Water until the compost is damp – not waterlogged.

GOOD ADVICE

- Drill holes in plastic troughs for drainage if none exist.
- Plant bulbs in rows or clusters, with the largest bulbs lowest. Cover with potting compost, water in and leave in a shady spot until late winter, keeping the soil moist. Then move to a sunny spot to encourage flowering.
- Add charcoal to prevent bulbs rotting.

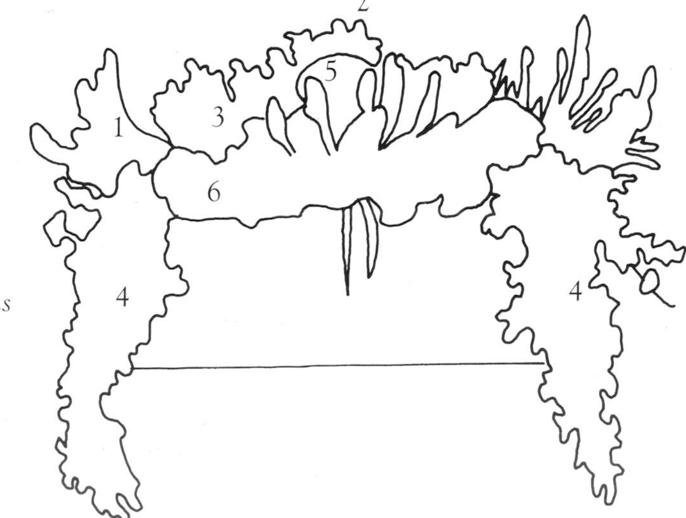

1 *grape hyacinths*
2 *box*
3 *rosemary*
4 *ivies*
5 *polyanthas*
6 *primulas*

Spring Hanging Basket

After the dull dark days of winter, it's cheering to see bulbs bursting into life, especially if you can position them in a basket near your front door.

If you intend using bulbs, you need to start planning in early autumn and planting in mid-autumn. Alternatively, visit a garden centre in early to mid-spring for a colourful display of ready-grown bulbs that are hardy enough to withstand frosts and strong winds.

Choose dwarf varieties for windowboxes where possible or you may find that the stems grow too long and the plants droop or become too lanky and make the basket look uneven.

YOU WILL NEED

a wicker basket, 50 cm (20 in) in diameter · plastic liner · moss · 12 daffodils (*Narcisssus* 'Tête-à-Tête') · 2 dwarf tulips · 1 cowslip (*Primula veris*) · 1 grape hyacinth (*Muscari*) · 1 scilla (*S. sibirica*) · 1 spurge (*Euphorbia*) · 2 primroses (*Primula vulgaris*) · 2 daisies (*Bellis*)

PERIOD OF INTEREST
February to April

LOCATION
The grape hyacinth and daisies will do better in a sunny spot.

FOR CONTINUITY
When the bulbs have flowered, opt for a pink, green and white theme by keeping the spurge and daisies and adding alpines such as *Phlox douglasii* and *Oxalis adenophylla*.

1 Gather your ingredients. Choose an interestingly shaped basket that is also deep enough to hold the compost and that has a very strong handle as the basket will be heavy when filled with mature plants and wet compost.

2 Line the basket with moss, then cut the plastic to shape for the liner. Make slits to ensure adequate drainage and fill to two-thirds with potting compost. Decide which is to be the front of your basket.

3 Start by placing the taller plants at the back. The cowslip, grape hyacinth and scilla will add central colour. Infill with spurge, primroses and daisies at the front. Top up the compost, firm and water in well.

GOOD ADVICE

- Do not position the basket in a windy corner as the compost will dry out.

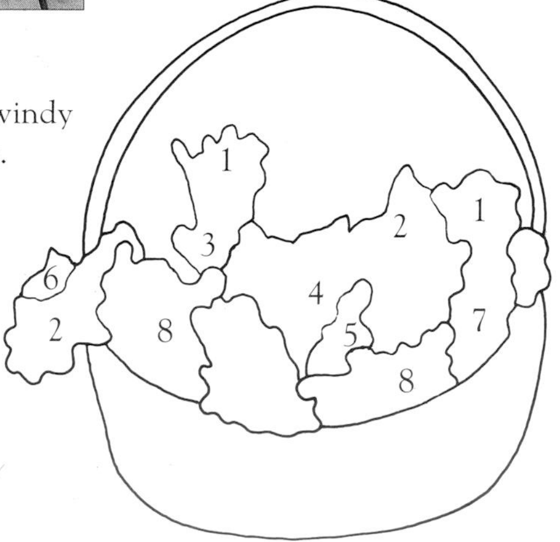

1 *daffodils*
2 *tulips*
3 *cowslip*
4 *grape hyacinth*
5 *scilla*
6 *spurge*
7 *primroses*
8 *daisies*

Planting a Windowbox

Just like hanging baskets, windowboxes have a tendency to dry out, so some of these steps are specifically intended to counteract this problem.

The planting scheme itself can be varied to suit almost any colours you have in mind – there are plenty of ideas throughout the book – but we thought these bright oranges and blues made a refreshing change from more traditional colours. Hydrangeas make particularly good container plants as long as their soil is ericaceous (that is, acidic not alkaline) and they are kept well watered.

YOU WILL NEED

a wooden trough, 65 cm (26 in) x 23 cm (9 in) · a hand-operated or electric drill · broken crocks or pebbles plastic for lining · ericaceous potting compost · water-retaining granules fertiliser granules · 1 hydrangea 1 scabious (*Scabiosa*) · 6 African marigolds (*Tagetes erecta*) · 8 French marigolds (*T. patula*) · 10 petunias

PERIOD OF INTEREST
June to October

LOCATION
In full sun to light shade. These vibrant colours will benefit from a single-colour background.

FOR CONTINUITY
Pinch back the petunias to encourage growth. In autumn, plant the hydrangea out in the garden and replace the scabious, petunias and marigolds with ericaceous-loving winter heathers.

1 Drill a drainage hole in the base every 15 cm (6 in). Cover the base inside with crocks or pebbles to prevent soil from leaching out. Then insert the plastic lining to prevent the wooden box bowing through damp.

2 Fill to one-third with potting compost (for most plants a multi-purpose variety is adequate but the hydrangea will need an ericaceous one) adding a few handfuls of water-retaining granules. Sprinkle in the fertiliser.

3 Test run the position of the plants while still in their pots. Then insert the hydrangea. Place the scabious on the left. Add the African and French marigolds and infill with petunias. Fill with compost and water in.

GOOD ADVICE

- Charcoal scattered in the base will prevent container plants from becoming mildewed through overwatering.

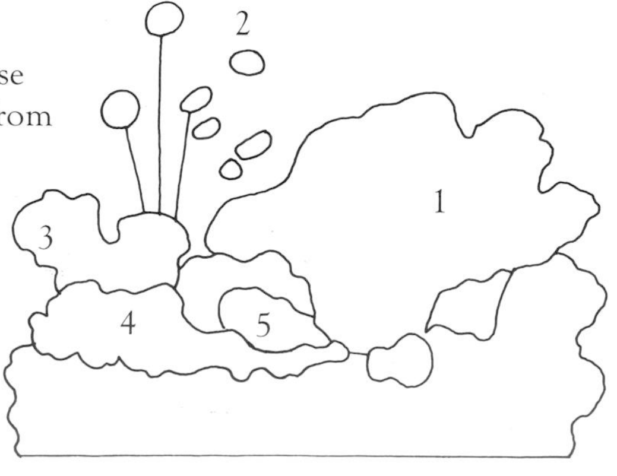

1 *hydrangea*
2 *scabious*
3 *African marigolds*
4 *French marigolds*
5 *petunias*

Planting a Hanging Basket

To recreate the colourful summer basket opposite, follow the step-by-step photographs and instructions. These are all popular plants for container gardening. However, should a particular plant be unavailable, your garden centre will advise you of a suitable substitute.

We have used quite a large basket as smaller versions hold too little compost and tend to dry out quickly. Lack of water is the most common reason for unsuccessful hanging baskets. A liner made of plastic, fibrous matting or foam will help to retain water. A 'self-watering' basket has a reservoir of water in its base.

YOU WILL NEED

a plastic-coated wire basket, 35 cm (14 in) in diameter · moss · fibrous, plastic or foam liner · potting compost · 1 trailing begonia · 1 *Plectostachys serphyllifolia* · 1 pelargonium · 1 pot trailing lobelia · 1 bidens (*B. aurea*)
1 pot verbena · 1 pink busy lizzie (*Impatiens*) · 3 purple petunias · 1 diascia

PERIOD OF INTEREST
June to September

LOCATION
In full sun to light shade. Leave plenty of room around this basket as the bidens will proliferate.

FOR CONTINUITY
Apply a liquid feed every 7–10 days. Deadhead the lobelia, diascia and bidens regularly.

1 First assemble your 'ingredients'. You may find it easier to put your basket on a bucket. Then plan the arrangement, with the tallest plants to the rear and the smallest, trailing plants around the edges or inserted from below.

2 Line the whole basket with moss. Add the fibrous, plastic or foam liner, cutting holes for lower planting if the liner allows. Half fill with compost. Remove plants from their pots with care, to avoid root damage.

3 Plant the begonia and plectostachys in the centre. Thread lobelia and bidens roots into the base from the outside. Add the other plants, then fill with compost to the brim. Firm the plants and water well.

1 *begonia*
2 *pelargonium*
3 *lobelias*
4 *bidens*
5 *verbena*
6 *busy lizzie*
7 *petunias*
8 *diascia*
9 *plectostachys*

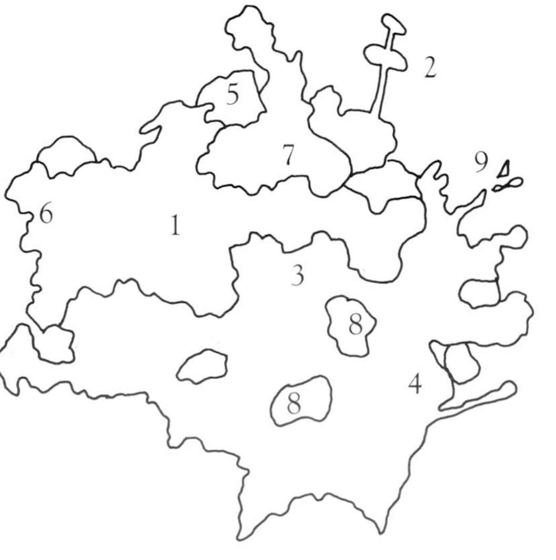

Planting a Wall-mounted Half-basket

Wall-mounted baskets give a plain wall life and colour. They take up less space than hanging baskets and generally hold fewer plants. Because less compost is needed, however, they are more prone to drying out and need careful attention and regular watering. Half-baskets also require strong, secure wall fixings, as they can be very heavy when full.

Take into consideration the aspect of your wall when choosing plants. If it is a shady spot, opt for busy lizzies or fuchsias. A hot spot could make an ideal site for nasturtiums, pelargoniums or other sun-loving plants.

YOU WILL NEED

a cast-iron or plastic-coated wire half-basket, 35 cm (14 in) in diameter · plastic liner · moss · fertiliser granules · 6 busy lizzies (*Impatiens*) · 6 lobelias · 1 trailing pelargonium · 1 upright pelargonium 1 variegated ground ivy (*Glechoma hederacea* 'Variegata') · 1 lysimachia (*L. lyssii*) · 1 red monkey musk (*Mimulus*) 1 purple brachycome

PERIOD OF INTEREST
June to September

LOCATION
In a sunny spot on a south- or west-facing wall.

FOR CONTINUITY
Underplant with bulbs in autumn and top with red primroses and ivy for the Christmas season.

1 Line the basket with black plastic, ensuring the back wall is well protected. Fill with moss in front of the plastic, with the greenest part outermost, right up and over the top edge. Fill one-third with potting compost.

2 Sprinkle in granular fertiliser and add a further third of compost. Starting at one side, make cuts through the plastic liner halfway down the basket and alternately insert busy lizzies and lobelia through the slits.

3 Test the rest of your plan by placing the plants in the basket still in their pots. Place the upright pelargonium and ground ivy in the centre and all trailing plants to the sides. Plant up the basket and firm and water the plants in.

GOOD ADVICE

- To prevent marks on your wall, line the whole of the basket with plastic sheeting, especially the back.

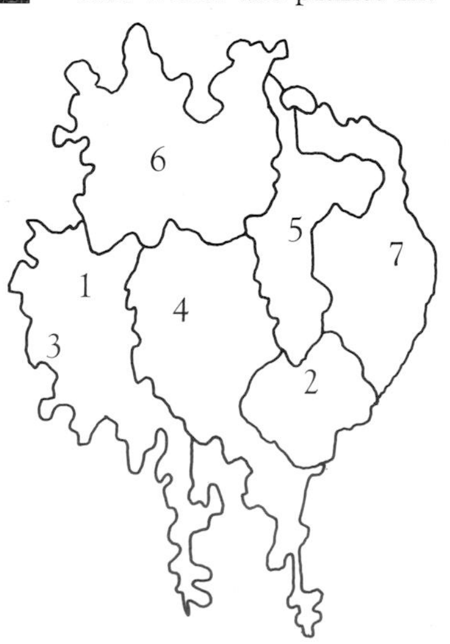

1 *busy lizzies*
2 *lobelias*
3 *pelargoniums*
4 *ground ivy*
5 *lysimachia*
6 *monkey musk*
7 *brachycome*

Country Cottage

Windowboxes are particularly evocative of a rural idyll. For many of us, a cottage with roses around the door will remain a dream, but windowboxes of fragrant lavender, stocks and chamomile are perfectly achievable.

Although known to many gardeners, snow-in-summer and creeping Jenny are not often used in containers. Their rampaging character will however be contained by a windowbox and they are both fairly hardy, but will not stand drying out. Monkey musk deserves a gold star for its long-standing versatility: flowering from late spring onwards, it produces rewarding blooms and trails prettily over the sides of the box.

YOU WILL NEED

a terracotta trough, 50 cm (20 in) x 23 cm (9 in) · 3 stocks (*Matthiola*)
1 dwarf lavender (*Lavandula angustifolia* 'Hidcote') · 1 chamomile (*Anthemis*)
1 snow-in-summer (*Cerastium tomentosum*)
1 creeping Jenny (*Lysimachia nummularia* 'Aurea') · 1 variegated ground ivy (*Glechoma hederacea* 'Variegata')
1 cream monkey musk (*Mimulus*)

PERIOD OF INTEREST
May to mid-August

LOCATION
In sun or light shade. This windowbox was on an east-facing wall.

FOR CONTINUITY
Replace the stocks with dwarf tobacco plants (*Nicotiana* x *sanderae*, Domino Series) available in various soft colours.

1 Check that your windowbox has drainage holes. Then cover these with a layer of crocks or pebbles and add the compost. Assemble the plants, with trailing ones to the fore and taller plants to the rear of the box.

2 Space the stocks out. Make sure the crown of the root is 2.5 cm (1 in) below the surface. Add lavender and chamomile between the stocks, then, along the front, snow-in-summer, creeping Jenny, ground ivy and monkey musk.

GOOD ADVICE

- Choose dwarf varieties of lavender or it will grow too tall and block the light.
- Deadhead stocks and they will bloom again.
- Age a terracotta windowbox by painting it with natural yoghurt and leaving it on the ground outside for a few weeks while nature's micro-organisms break down the yoghurt to create an interesting 'look'.
- Choose more unusual trailing plants, such as monkey musk and campanula, for a country look.

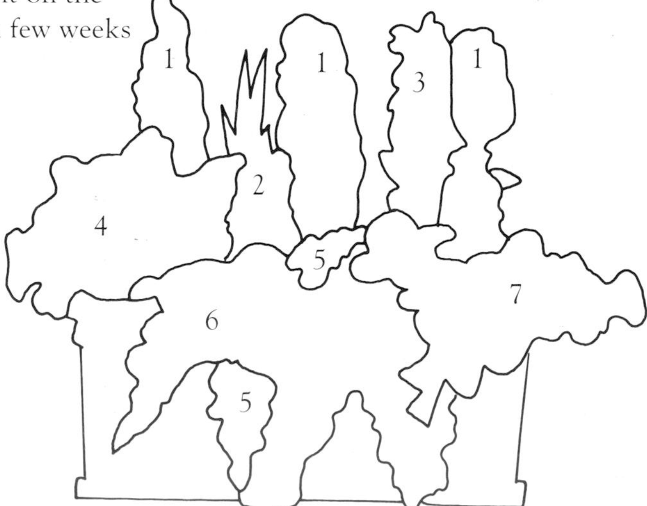

1 *stocks*
2 *lavender*
3 *chamomile*
4 *snow-in-summer*
5 *creeping Jenny*
6 *ground ivy*
7 *monkey musk*

Wildflower Windowbox

A windowbox of wildflowers makes an unusual contrast to bold displays of well-known plants. These wildflowers are suitable for any container.

Flowering from late spring to mid-summer

Yellow: rock rose, birdsfoot trefoil, horseshoe vetch, silverweed, monkey musk, feverfew

Blue: ivy-leaved toadflax, cornflower, heartsease, meadow cranesbill, bellflowers, wild thyme, wild marjoram

Pink/Red: herb Robert, pheasant's eye, poppy, mallow, dusky cranesbill, columbine

White: barren strawberry

Flowering from late summer to mid-autumn

Yellow: toadflax, creeping Jenny, orange hawkweed

Blue: small scabious, bellflower, harebell

YOU WILL NEED

a wooden windowbox, 60 cm (24 in) x 25 cm (10 in) · water-retaining granules
1 columbine (*Aquilegia vulgaris*)
1 cornflower (*Centaurea*) · 1 dwarf mallow (*Malva*) · 1 feverfew (*Tanacetum parthenium*) · 1 oriental poppy (*Papaver orientale*) · 1 silver dead nettle (*Lamium maculatum* 'Beacon Silver')

PERIOD OF INTEREST
April to July

LOCATION
In gentle sun to light shade.

FOR CONTINUITY
Replace the poppy with toadflax or herb Robert in mid-summer.

1 Ensure that the box has drainage holes, then crock the base. Fill with potting compost, adding slow-release fertiliser half way. Water-retaining granules are a good idea as many wildflowers thrive in damp conditions.

2 Plant the columbine at one end and the cornflower at the other. Infill with the mallow, feverfew and poppy, allowing the silver dead nettle to spill over the front of the box. Top up with compost and water in well.

GOOD ADVICE

- Trim shabby growth.
- Wildflowers often thrive in semi-shady spots.
- Keep the compost moist; moisture-retaining granules should be added when filling with compost.
- Provide adequate drainage by putting broken crocks or pebbles in the base.
- An asymmetrical arrangement can look particularly attractive. Plan your scheme to have larger plants at one end with perhaps a trailing plant at the other.
- Some garden centres have special wildflower areas – most plants are now clearly labelled but it is worth making enquiries if you need to check the compatibility of certain plants.

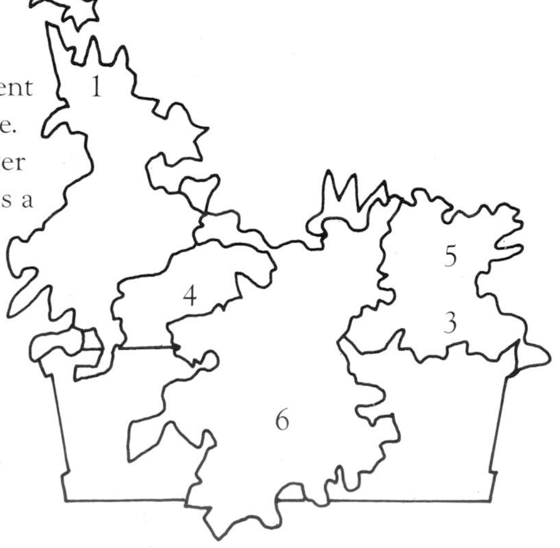

1 *columbine*
2 *cornflower*
3 *mallow*
4 *feverfew*
5 *poppy*
6 *dead nettle*

Mixed Hedgerow

A wildflower garden need not necessarily be a rambling meadow. A wildflower basket can provide you with sturdy flowers you recognise from the hedgerows of your childhood, with delicate blooms and fascinating foliage, all to be enjoyed as a compact sphere.

The one disadvantage of wildflowers is their brief flowering season, but this can be overcome by planting so that at any one time some flowers are in bud while other blooms are fading. On page 90 we have listed wildflowers for baskets and windowboxes and the flowering seasons.

Your local garden centre is likely to have a wildflower section and you should be able to order other varieties through them. You should never take plants from the wild!

YOU WILL NEED

a wire basket, 35 cm (14 in) in diameter
moss · 1 pink dusky cranesbill (*Geranium phaeum*) · 3 rock roses (*Helianthemum*)
1 blue wild geranium (*G.* 'Johnson's Blue')
2 periwinkles (*Vinca*): 1 plain-leaved, 1 variegated · 1 bellflower (*Campanula*)
1 soapwort (*Saponaria*)

PERIOD OF INTEREST
May to July

LOCATION
In sun, at eye level, where the delicate flowers can be appreciated.

FOR CONTINUITY
As they fade, replace the rock roses with creeping Jenny and chamomile.

1 Balance the basket on a pot or bucket, then line the basket with damp moss, green side outermost. Place a liner of plastic, foam or fibre in the base, then fill with compost, adding slow-release fertiliser half way.

2 Plant the cranesbill and rock roses in the centre with the other plants at the sides. Guide these beneath the top wire to encourage them to trail down the sides and cover the base. Top up with compost and water well.

GOOD ADVICE

- Find the wildflower corner of your local garden centre and study the plant labels.
- A wildflower leaflet is sometimes available too.
- Choose flowers that bloom in the same season where possible.
- Plan your colour scheme in advance, or opt for a random assortment as here.
- Keep the basket well-watered and do not allow the compost to dry out.
- For a random ball arrangement, thread some plants through the sides and base.

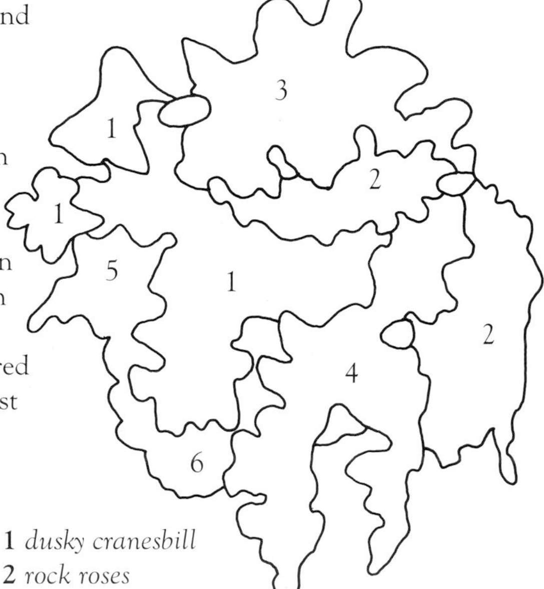

1 *dusky cranesbill*
2 *rock roses*
3 *wild geranium*
4 *periwinkles*
5 *bellflower*
6 *soapwort*

Evening Fragrance

Many plants with scented leaves need to be brushed past to release their perfume but those with blooms often release their fragrance in the evening.

The tobacco plants used here will produce blooms from mid-summer until the early autumn frosts. They can grow up to 30 cm (12 in) tall and tend to lean forward and trail prettily over the windowbox.

Some windowboxes, such as this one, are fixed to the wall and will need to be planted in situ. Others have a plastic liner, which means that you can plant on a suitable surface and then slide the liner into the windowbox.

YOU WILL NEED

a wooden windowbox, 50 cm (20 in) x 23 cm (9 in) · 1 lemon-scented pelargonium · 4 dwarf lavender (*Lavandula angustifolia* 'Hidcote') 1 tray mixed tobacco plants (*Nicotiana* x *sanderae*, Domino Series) · 1 tray white tobacco plants (*Nicotiana alata* 'Dwarf White Bedder')

PERIOD OF INTEREST
July to October

LOCATION
In sun or semi-shade near a doorway or window so that you can enjoy their mingled fragrance.

FOR CONTINUITY
For autumn colour and contrasting scent, replace with dwarf perennial coneflowers and miniature chrysanthemums.

1 Check that your windowbox has drainage holes. You may need to drill some. Then line the box with slit plastic and fill with potting compost. Plan your planting scheme before you plant up the windowbox.

2 At this stage you could scatter some slow-release fertiliser half way down the box. Plant the pelargonium so that it will trail over the front, then infill with the lavender. Plant the taller tobacco plants to the rear. Water well.

GOOD ADVICE

- Choose dwarf varieties of lavender and tobacco plants as some are fairly tall.
- Check that the tobacco plants you have chosen are fragrant – some varieties are more heavily scented than others.
- You can buy liners for wall-mounted windowboxes, which make planting up far easier.
- Place your box where it will be brushed by people passing by, otherwise the scent will be released in only small quantities.

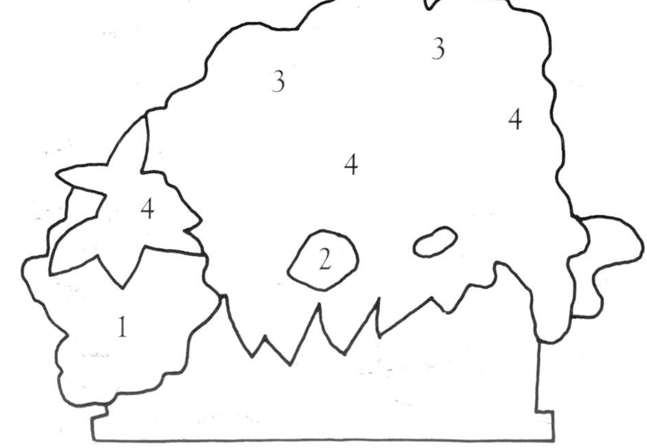

1 *pelargonium*
2 *lavender*
3 *mixed tobacco plants*
4 *white tobacco plants*

Scented Basket

Hang a scented basket where you will brush against the leaves as you walk by, releasing their fragrance. Baskets of silvery plectostachys can look most effective, and its tiny leaves release a wonderful aroma when you rub them between your fingers.

Pineapple mint, like all mint, can be invasive but it is worth including for its pleasant, fruity fragrance when the leaves are gently crushed – hence its name. To make sure the mint does not spread too far, you can contain root growth by inserting vertical slates when planting, or by simply keeping it in its pot.

The pinks and heliotrope look good together and have a delicate perfume. The violets are included not for their scent but to complement the general colour scheme.

YOU WILL NEED

a wooden basket, 40 cm (16 in) x 30 cm (12 in) · plastic liner · 8 pinks (*Dianthus* 'Doris Allwood') · 1 heliotrope 1 *Plectostachys serphyllifolia* · 1 pineapple mint (*Mentha suaveolens* 'Variegata') · 1 pot white ivy-leaved violets (*Viola hederacea*)

PERIOD OF INTEREST
June to September

LOCATION
In sun, where these gentle scents will be appreciated.

FOR CONTINUITY
Replace fading plants with citrus-smelling *Houttuynia cordata* 'Chameleon' and a trailing fuchsia, *F.* 'Pink Galore'.

1 If your basket is wooden, you will need to drill holes in the base first. Having lined it with slit plastic, cover the base with pebbles. Fill loosely with potting compost. Then plant the pinks centrally within the basket.

2 Add the heliotrope, plectostachys, pineapple mint and violet, one plant in each of the four corners. Top up with potting compost, then firm gently. Water the plants in and hang the basket in its final position.

GOOD ADVICE

- Use a wood preservative to prolong the life of your basket. Leave it to dry for a day.
- Line it with plastic to protect the timber.
- Cut slits for drainage.
- Pinks need full sun in order to flower so keep your basket in a sunny spot.
- Deadhead the plants regularly.
- Stop the mint from spreading by keeping it within its pot or cornering it with slates.

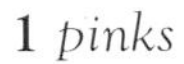

1 *pinks*
2 *heliotrope*
3 *plectostachys*
4 *pineapple mint*
5 *violets*

Windowbox for a Shady Area

Few of us would plan a windowbox for shade without being prompted but it can be particularly rewarding. You may not have a box overflowing with flowers, but many shade-loving plants have graceful lines, strong colours and textures that work well together.

Some plants do flower in shade, of course. You could use busy lizzies or fuchsias. We chose to use bedding begonias with their lustrous, dark leaves. Hardy dead nettle is also surprisingly attractive with its yellow or pink flowers.

Shade-loving plants appreciate moist conditions so include water-retaining granules when planting up.

YOU WILL NEED

a wooden windowbox, 60 cm (2 ft) · plastic or other liner · water-retaining granules
1 Boston fern (*Nephrolepis exaltata* 'Bostoniensis') · 1 yellow dead nettle (*Lamium galeobdolon* 'Hermann's Pride')
2 hostas (*H. fortunei* 'Aureomarginata')
3 bedding begonias (*B. semperflorens*)
1 *Fuchsia* 'Autumnale'

PERIOD OF INTEREST
June to September

LOCATION
In semi-shade. Sunshine for part of the day will give better leaf colour and flowers.

FOR CONTINUITY
Replace plants after the first frosts with deep orange *Cotoneaster franchetii* and deeply lobed, variegated ivies.

1 Check that there are drainage holes in the windowbox, and drill some if there are not. Line the base with plastic or other material. Then spread out crocks or pebbles and fill to half way up the box with potting compost.

2 Add water-retaining granules and slow-release fertiliser. While they are still in their pots, plan how your plants are to look: where you would like trailing foliage, height and colour. Move them until you are happy.

3 Insert the largest plant – the fern – and then the dead nettle at the other end of the box. Infill with hostas and begonias, with fuchsia trailing over the front. Top up the compost, then firm and water in.

GOOD ADVICE

- Treat wooden boxes with wood preservative and allow to air for a day.
- Snip off browning fronds or leaves.

1 *fern*
2 *dead nettle*
3 *hostas*
4 *begonias*
5 *fuchsia*

Shady Surprise

Most hanging baskets thrive in sun but many plants also bloom well in semi-shade, so shady corners should not be forgotten. Fuchsias, in particular, are known for their partiality to shade, and busy lizzies, too, will flower without sunshine as long as they are placed in the shade only after the first flowers have opened.

Some gardeners are surprised to learn that begonias will flower in partial shade. Pansies are known as winter-flowering plants so it is not unusual for them to bloom in a shady corner.

The alternative is to choose plants that have particularly attractive foliage – ferns and hostas are an obvious choice. These need damp soil and moist conditions, so frequent spraying and water-retaining granules are important.

YOU WILL NEED

a metal basket, 35 cm (14 in) in diameter
moss · plastic or other liner · water-retaining granules · 1 semi-trailing fuchsia (*F.* 'Orange Crystal') · 2 busy lizzies (*Impatiens*) · 1 monkey musk (*Mimulus*)
1 coral begonia (*B.* 'Chanson')
1 *Convolvulus sabatius*

PERIOD OF INTEREST
June to September

LOCATION
Semi-shade would suit these moisture-loving plants.

FOR CONTINUITY
In autumn introduce a mix of violets and pansies in shades of white, lilac and purple with a splash of yellow.

1 Gather your ingredients. Place the basket on a bucket to prevent it from rocking, then line it with damp moss, green side outermost. Line with plastic, then fill it with compost and water-retaining granules.

2 Plant the fuchsia, one busy lizzie and monkey musk on three sides of the basket with the upright begonia in the centre. Trail the convolvulus beneath the top wire. Plant the other busy lizzie through the sides. Water in well.

GOOD ADVICE

- Avoid splashing water on the leaves of the begonia as this can cause rotting.
- Deadhead regularly and feed every two weeks with a liquid fertiliser.
- Shade-loving plants generally appreciate lots of water, so keep the compost moist. Fuchsias and begonias both need plenty of water.
- Take a closer look at the strongly defined foliage plants, including most ferns. They can look surprisingly attractive, particularly if your windowbox is painted a bold colour.
- Many winter-flowering plants, such as pansies, will flower in shady spots.

1 *fuchsia*
2 *busy lizzies*
3 *monkey musk*
4 *begonia*
5 *convolvulus*

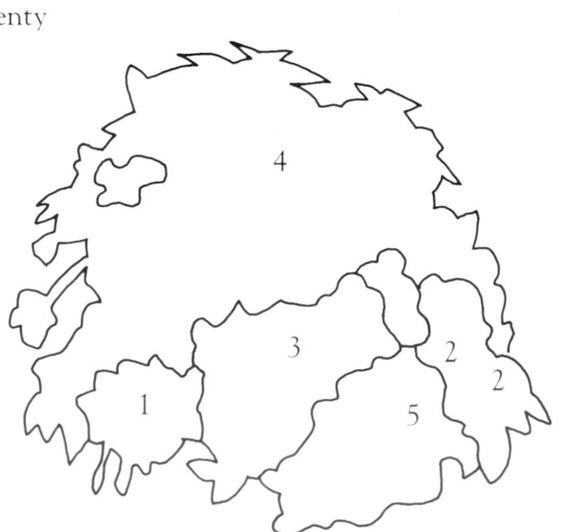

Alpine Appeal

The small size and delicate appearance of alpines often belies their hardiness. Used to surviving in mountainous regions, they can withstand dry periods but dislike being overwatered. Terracotta is therefore a particularly suitable material for an alpine windowbox as any excess moisture will drain away naturally through the porous walls and base of the box. (You will still need drainage holes, of course.)

You may well feel tempted by other alpines when faced with a display at your garden centre or nursery. Other favourites not included here are: rock roses, sedums, houseleeks and lewisia. Aim to mix flowering and foliage plants and to group complementary colours.

YOU WILL NEED

a reconstituted-stone trough, 45 cm (18 in) x 30 cm (12 in) · well-drained potting mix of half fine horticultural grit: half potting compost · 1 bellflower (*Campanula carpatica*) · 1 creeping Jenny (*Lysimachia nummularia* 'Aurea') · 1 thrift (*Armeria maritima* 'Dusseldorf Pride') 1 saxifrage · 1 *Chrysanthemum hosmariense* 1 campion (*Silene schafta* 'Robusta')

PERIOD OF INTEREST
May to July

LOCATION
In sun, on a windowsill, bench or steps, where they cannot be overlooked.

FOR CONTINUITY
Infill with tumbling ted (*Saponaria ocymoides*) for later summer colour.

1 Line the base of the trough with a thick layer of crocks or pebbles. Then mix 50:50 grit and compost and fill the trough. The rooting system of alpines is very shallow so they will not need deep compost.

2 Tap the plants from their pots, then gently tease out the root system. Starting at one end, work along the trough, inserting one plant to the front and the next to the rear; a large stone in the middle looks effective.

3 Water in carefully, avoiding the crowns of the plants and the tops of leaves as alpines do not like to be waterlogged. If there are spaces to fill, you can add other suitable plants later on.

GOOD ADVICE

- To create well-drained soil for alpines, mix equal parts of fine horticultural grit and potting compost.

1 *bellflower*
2 *creeping Jenny*
3 *thrift*
4 *campion*
5 *chrysanthemum*
6 *saxifrage*

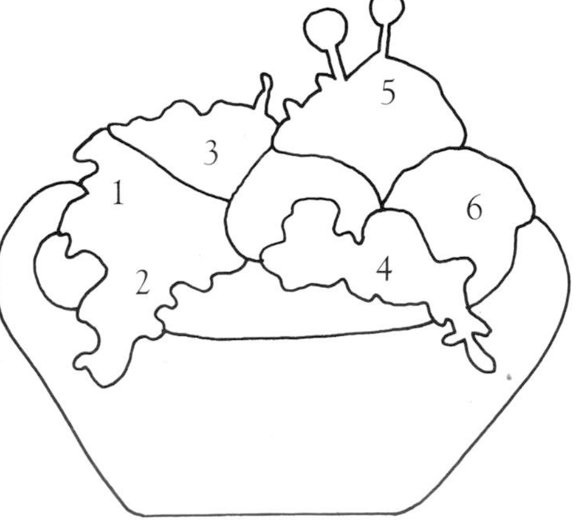

Floral Feast

Alpines prefer dryish soil, so water sparingly and, if you have a particularly rainy week, shelter the plants under cover for a while. Because alpines hate having wet roots, we used a thin fibre liner through which moisture can pass. It also has planting slits at lower levels so that plants can cover the basket base.

This basket was planted in early spring when most of the plants were very small and relatively inexpensive. It is composed of plants that will flower in groups from early spring to early autumn. Hung from a post of climbing honeysuckle, it created an exceptionally pretty corner beside a doorway.

YOU WILL NEED

a plastic-coated metal basket, 38 cm (15 in) in diameter · well-drained potting mix of half fine horticultural grit: half potting compost · 1 rock rose (*Helianthemum* 'The Bride') · 1 pink (*Dianthus* 'Dewdrop') 1 *Sisyrinchium* 'Mrs Spivey' · 3 bloody cranesbill (*Geranium sanguineum* var. *striatum*) · 3 *Gypsophila repens* 'Dubia'. 1 speedwell (*Veronica pectinata*) 1 tumbling ted (*Saponaria ocymoides*) 1 bellflower (*Campanula* 'Birch Hybrid') 1 *Sedum spurium* var. *album* 1 *Diascia* 'Ruby Field'

PERIOD OF INTEREST
May to September

LOCATION
In sun; shelter from heavy rain.

FOR CONTINUITY
Trim back dead growth and feed every two weeks.

1 Gather your ingredients. Plan to have trailing plants at the sides and the more upright rock rose, pink and sisyrinchium in the centre. Insert the liner; add the gritty compost. Plant the rock rose, pink and sisyrinchium.

2 Plant the cranesbill and gypsophila in clumps to form larger plants later. Encircle the pink with trailing plants, planting from below and through the slits in the liner wherever possible. Top up the compost and water in gently.

GOOD ADVICE

- Alpines like well-drained soil so mix half fine horticultural grit and half multi-purpose potting compost in a bucket before filling your basket.
- Many alpines dislike having the top of their rootball (the crown) watered. Aim to water in between plants where possible.
- Make slits in fibre liners for planting through the sides and base to ensure an evenly covered hanging basket.
- Choose plants that flower in succession whenever possible.

1 *rock rose*
2 *pink*
3 *sisyrinchium*
4 *cranesbill*
5 *gypsophila*
6 *speedwell*
7 *tumbling ted*
8 *bellflower*
9 *sedum*
10 *diascia*

Cheap and Cheerful

It is easy to buy plants in season and create impressive windowboxes and baskets in the space of an hour, but this can prove expensive. Many plants can be grown from seed or cuttings and others bought to infill, if necessary, when your basket or box is ready to be planted up.

Fuchsias, pelargoniums, *Petunia* 'Surfinia', verbenas and busy lizzies are all easy to grow from cuttings. You can simply stand the cutting in a glass of water for a couple of weeks until several white roots have grown. The more professional method, however, is to cut the stem straight across, dip it in hormone rooting powder and insert it in a covered container filled with proprietary cuttings compost.

YOU WILL NEED

a plastic-coated wire basket, 35 cm (14 in) in diameter · foam liner · 1 packet lobelia seed · 1 packet *Bidens aurea* seed
2 *Fuchsia* 'Blowick' · 2 *Petunia* 'Surfinia'
6 busy lizzies (*Impatiens*)
2 *Verbena* 'Sissinghurst'

PERIOD OF INTEREST
June to September

LOCATION
In sun for much of the day, ideally on a west-facing wall. A south-facing wall may be too drying and will mean extra watering.

FOR CONTINUITY
Keep moist and well fed to ensure a good display all summer long.

1 Sow the lobelia and bidens thinly in trays of seed compost in early spring. Water with a fine spray. Cover and keep warm until the first leaves appear. Then remove the cover. Transplant the seedlings once 5 cm (2 in) tall.

2 Fuchsias, petunias, busy lizzies and verbenas can be grown from cuttings. Line the basket with foam. Plant the fuchsias in the centre, angling the other plants to trail through. Firm the plants gently, then water them well.

GOOD ADVICE

- For a good summer-long display choose perennial lobelia plug plants such as *L. richardsonii* and 'Azureo'.
- Use a saucer and knife to help to control the position of the seeds on the compost.
- Cover the seeds with cling film with holes pierced in it to prevent too much condensation. When the first leaves appear, lift the cling film each day to 'air' the plants.
- Leave the seed tray in a warm, light but not too bright place to allow the seeds to germinate. If it seems very bright, lay a sheet of newspaper over the tray.
- When transplanting, hold the seedlings by their top leaves, not by the roots. Use a pencil to make holes in the compost.

1 *lobelia*
2 *bidens*
3 *fuchsias*
4 *petunias*
5 *busy lizzies*
6 *verbenas*

Home Grown

You can achieve a spectacular windowbox with a very low financial outlay. Most of these plants were grown from cuttings and the trailing lobelia and verbena from seed. Growing from seed is easier if you have a greenhouse, but seeds and cuttings will also grow happily on a warm windowsill.

Start the seeds off in a tray of moist seed compost set in a warm place for ten days or so. Cover the tray with cling film with tiny holes in it. Ready-made windowbox gardening kits are now available with narrow seed boxes and raised, clear-plastic lids with air holes. Once the seeds have germinated and the first leaves appear, take off the cover and move the tray to a sunny position.

YOU WILL NEED

a plastic trough, 90 cm (36 in) x 20 cm (8 in) · 1 packet lobelia seeds
1 packet verbena seeds · 2 zonal pelargoniums · 5 busy lizzies (*Impatiens walleriana*) · 1 bedding begonia
1 pendulous begonia · 1 double white chamomile (*Chamaemelum*)

PERIOD OF INTEREST
June to September

LOCATION
In sun, on a west-facing sill where there is room for the plants to trail.

FOR CONTINUITY
Water twice daily if hot. Transplant dwarf China asters such as *Callistephus chinensis* 'Colour Carpet' from the garden or buy pot-grown plants for a late summer display.

1 Take cuttings of pelargoniums, busy lizzies, begonias and chamomile before the first frosts arrive, planting them in pots after four weeks or so. Sow the lobelia and verbena in early spring; transplant a month later.

2 Plant up your windowbox at the end of spring. Fill with crocks, then compost. Plant the pelargoniums, busy lizzies and begonias, then infill with chamomile and other trailing plants. Water them well.

GOOD ADVICE

- Mix tiny seeds with very dry silver sand to make sowing easier.
- Cover germinating seeds with cling film to keep the compost moist or with newspaper to keep out very bright light.
- Root cuttings in a mixture of equal parts peat and sharp sand.
- If seedlings start to become leggy in their search for light, turn them regularly, keep them in a light position and move them somewhere a little cooler.
- Make a note of planting times and how your seeds progress, for reference in future years.

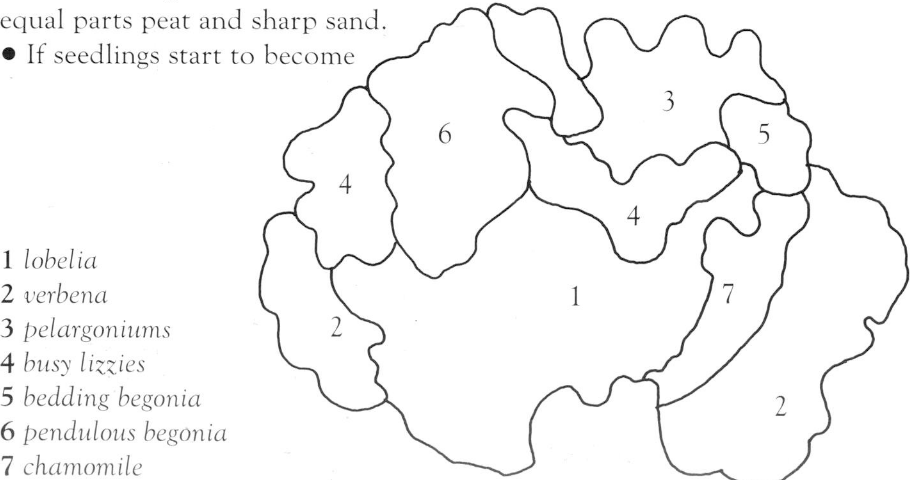

1 *lobelia*
2 *verbena*
3 *pelargoniums*
4 *busy lizzies*
5 *bedding begonia*
6 *pendulous begonia*
7 *chamomile*

Hanging Basket for Children

It is not always possible to find a suitable space for your child to garden nor to keep a child's garden contained within a set area. This is where baskets and windowboxes prove rewarding.

Choose bright colours and easy-to-grow plants that are not likely to disappoint. Petunias and marigolds are an obvious choice, as are busy lizzies and lobelia. Children will also enjoy planting up the baskets and boxes, and to them daily watering is a pleasure.

Results are relatively speedy from windowboxes and hanging baskets. Rather than growing from seed, use plants of a reasonable size so that growing time is comparatively short. A simple basket could be the start of an interest in gardening for years to come.

YOU WILL NEED
a wire basket, 25 cm (10 in) in diameter
moss · plastic liner · 2 *Petunia* 'Red Stripe' · 6 busy lizzies (*Impatiens*)
1 *Plectostachys serphyllifolia*
2 *Verbena* 'Red Cascade'

PERIOD OF INTEREST
June to September

LOCATION
In sun, at a child's head height or use a rise-and-fall hanging system.

FOR CONTINUITY
Pinch back the petunias when the stems are 15–20 cm (6–8 in) long to ensure bushy growth. Deadhead regularly and feed with liquid fertiliser.

1 Let the child line the basket with moss, green side outermost. Then line it with slit plastic. Half fill the basket with compost and plant the petunias in the top. An attractive alternative would be a trailing petunia hybrid.

2 With some adult help, plant the busy lizzies at intervals through the sides of the basket so that they will eventually almost cover the base. Trail the plectostachys and verbena below the top line of the basket.

3 Top up with compost and water in well. Allow the child the responsibility of daily watering and deadheading blooms so that they can appreciate the results of their first attempt and may try other arrangements.

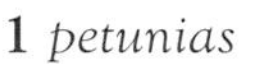

1 *petunias*
2 *busy lizzies*
3 *plectostachys*
4 *verbena*

Flowers of Childhood

Yellow marigolds and orange nasturtiums seem made for children as these easy-to-grow summer annuals make a splash of colour.

Children like to be part of every stage in gardening, from selecting plants (you will need to guide them towards compact, bushy plants) to planting and watering. Should you notice any pending disasters, you can do a little surreptitious gardening later on.

If you discuss the heights to which the plants will grow, the children will see the logic of planting them in descending order, with the trailing plants at the front. You could also explain how root systems work and how plants use chlorophyll to turn sunlight into food.

YOU WILL NEED

a plastic wall-mounted half-basket, 25 cm (10 in) in diameter · 1 marguerite (*Argyranthemum*) · 3 snapdragons (*Antirrhinum*) · 1 pot marigold (*Calendula*) 3 French marigolds (*Tagetes patula*) 2 nasturtiums (*Tropaeolum*) · 3 lobelia 2 brachycome

PERIOD OF INTEREST
June to September

LOCATION
In sun, at the window of a playhouse or in an easily accessible part of the house or garden.

FOR CONTINUITY
Prune out any dead plants; the others will soon fill the gaps.

1 Make drainage holes in the base of the windowbox if there are none. Add a few handfuls of gravel for drainage then half fill with potting compost. Set out the plants according to your plan and water them well.

2 Plant the marguerite, snapdragons and pot marigold to the rear, the French marigolds and nasturtiums in the centre, and the lobelia at the front. Arrange the brachycome to trail over one side. Top up with compost; water well.

GOOD ADVICE

- Allow the children to choose the colours of plants for themselves if possible. Bright colours are likely to be most popular.
- Choose fairly reliable plants, such as marigolds, lobelia and petunias, and aim to keep an eye on them yourself.
- Let the children plant the basket or windowbox and take care of it by watering, feeding and deadheading carefully with 'safe' scissors.
- Small baskets tend to dry out quickly so water frequently.
- Be very careful that any plants you use are not poisonous.

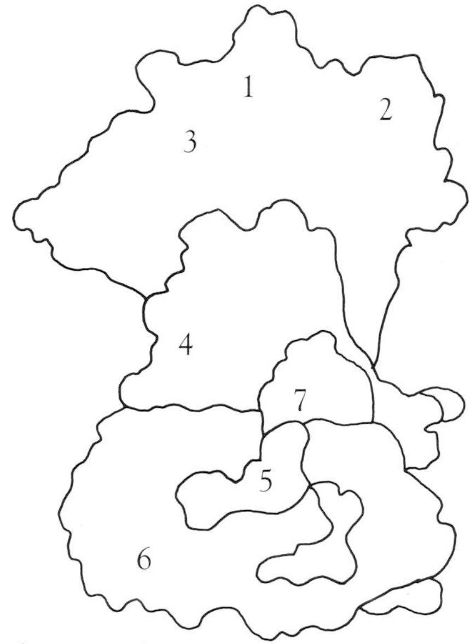

1 *marguerite*
2 *snapdragons*
3 *pot marigold*
4 *French marigolds*
5 *nasturtiums*
6 *lobelia*
7 *brachycome*

Edible Treat

The sometimes pungent scent given off by herbs can remind one of the Mediterranean, from where many culinary herbs originate. Perhaps because of their origins, most herbs prefer a sunny spot and gritty, free-draining soil. While a traditional herb garden can take up a great deal of space in a valuable sunny area of the garden, a windowbox or hanging basket occupies very little room – and the foliage can look highly attractive when a selection of herbs are planted together.

For a small and stylish windowbox that you can keep close at hand on a kitchen windowsill, a simple metal container such as this is ideal. Most herbs need a sunny situation to do well and this box would do equally well outside if conditions are right.

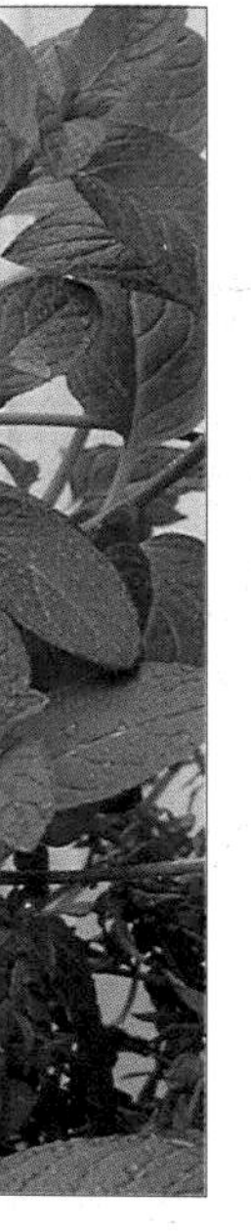

YOU WILL NEED

a metal container, 50 cm (20 in) · crocks well-drained potting mix of half grit: half potting compost · 1 lemon balm (*Melissa officinalis*) · 1 parsley (*Petroselinum*) 1 chive (*Allium schoenoprasum*) · 1 thyme (*Thymus*) · 1 mint (*Mentha*)

PERIOD OF INTEREST
May to September

LOCATION
In sun, on a windowsill or in a hot spot by the kitchen door for convenient snipping when preparing a meal

FOR CONTINUITY
Feed with liquid fertiliser every two weeks and do not overwater.

1 Test out the position of the plants, keeping them in their pots, to judge where you wish them to trail and how the leaf colours work best together. Place the crocks in the base of the container and fill with gritty compost.

2 Plant the herbs, teasing the roots out gently to help them to acclimatise quickly. Train them over the sides of the container if you wish. Top up the compost. Then firm and water in, taking care not to overwater.

GOOD ADVICE

- Cut herbs often to promote bushiness.
- If you are using the windowbox inside, you will not need drainage holes but you will have to be careful when watering to ensure that the soil does not become waterlogged.
- Remove flowerheads from chives etc. to promote leaf growth.

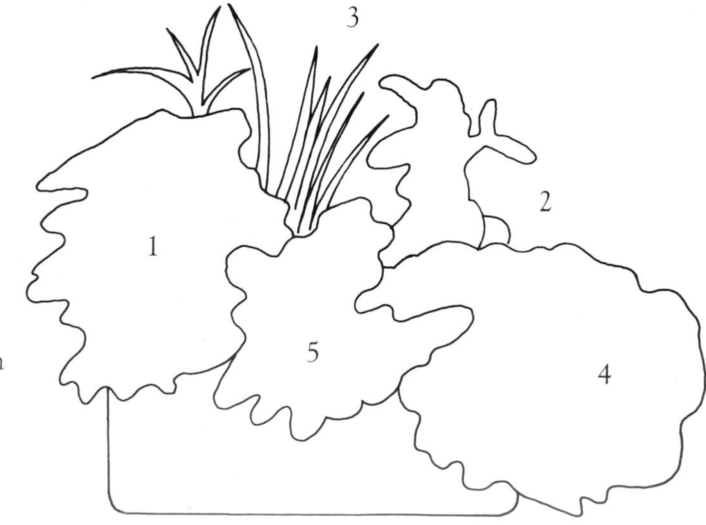

1 *lemon balm*
2 *parsley*
3 *chive*
4 *thyme*
5 *mint*

Herbal Basket

Here are a few ideas of ways to enjoy all these marvellous herbs that will be flourishing in your hanging basket.
Chives: best used at the end of cooking. Snip over soups and sauces, mix with cream cheese over jacket potatoes.
Fennel: use the aniseed-tasting stalk and leaves in tomato and fish sauces.
Mint: mostly used for mint sauce and with new potatoes. Also in tea.
Parsley: useful as a garnish, in stocks and to freshen breath. Useful in sauces for fish, ham and eggs. Freezes well.
Sorrel: a few handfuls (40 g/1½ oz) of the leaves are enough for a delicious soup. Tastes like spinach.
Tarragon: add a few sprigs to bearnaise sauce or wine vinegar for salad dressings.
Thyme: chopped in salad dressings or in stocks and casseroles. Dries well.

YOU WILL NEED

a wicker basket, 30 cm (12 in) in diameter
plastic liner · gritty potting mix · 2 fennel (*Foeniculum vulgare*) · 1 sorrel (*Rumex acetosa*) · 1 parsley (*Petroselinum*) · 1 mint (*Mentha*) · 1 thyme (*Thymus*) · 1 chive (*Allium schoenoprasum*) · 1 French tarragon (*Artemisia dracunculus*) · 1 sage (*Salvia*)

PERIOD OF INTEREST
May to September

LOCATION
In a sunny place where you may brush past and release the aromatic fragrances.

FOR CONTINUITY
Replace the fennel with a bright, ivy-leaved pelargonium such as *P.* 'Harlequin Alpine Glow'.

1 A plastic-lined wicker basket is ideal for herbs as long as you make a few slits in the plastic before filling. A ratio of about half grit and half compost gives the sort of free-draining environment that herbs prefer.

2 Plant the taller fennel and sorrel in the centre to the rear. Then add the mint, sage, parsley, thyme, chives and tarragon around the edges. Keep the compost moist but not soggy; the grit should keep the compost well-drained.

GOOD ADVICE

- Always ensure that herbs enjoy a hot and sunny spot.
- Feed weekly during the summer months.
- Most plants will die off and need replacing next spring.
- You could keep your herb basket hanging in the kitchen once it is well-established. Herbs enjoy sunshine though, so hang it outside occasionally in a sunny spot.
- Do not worry if the herbs dry out a little; they will revive if trimmed and watered.
- Fennel grows to about 60 cm (2 ft) tall so plant it only if you have enough room.

1 *fennel*
2 *sorrel*
3 *mint*
4 *sage*
5 *parsley*
6 *thyme*
7 *chives*
8 *French tarragon*

Tomato and Lettuce Windowbox

As lettuce hearts up in around ten days, you will find you can pick them and replace them with smaller plants while waiting for the tomatoes to grow and ripen – around six to ten weeks, depending on the amount of light and warmth they receive. Cos lettuce is a useful, upright variety for infilling between tomato plants.

Tomatoes are prone to aphids and may need spraying occasionally. Also, ensure that they have enough water so they do not wilt. 'Totem' is a compact, upright variety that will not grow too tall, but you will need to pinch out all side shoots. Alternatively plant bushy trailing 'Tumbler' tomatoes.

YOU WILL NEED

a reconstituted-stone planter, 60 cm (2 ft) · water-retaining granules
slow-release fertiliser
2 tomatoes 'Tumbler' or 'Totem'
3 cos lettuces

PERIOD OF INTEREST
July to September

LOCATION
In sun, on a sheltered balcony or in a glazed porch or conservatory for the most rapid growth and best crops.

FOR CONTINUITY
Replace any cropped lettuce with seedlings growing in small pots. Cut-and-come-again varieties such as 'Lollo Rosso' can be trimmed repeatedly and so avoid creating a gap.

1 Three lettuce and two tomato plants will give a good yield. Drill holes in the base of your windowbox if there are none already. Line the base with crocks or pebbles for drainage, then half fill with potting compost.

2 Scatter slow-release fertiliser and water-retaining granules at this level. Then fill to the brim with compost. Teasing their roots out gently, plant the tomatoes, and insert the lettuces at each side and in the centre.

3 Water in well. Make sure the compost covers the roots well. firm it down a little and add more if necessary. Do not crowd the box with more plants as they require a lot of nutrients to produce a good crop.

GOOD ADVICE

- Tomatoes need regular feeding and watering. Feed weekly with a liquid tomato fertiliser. Water twice daily in warm or windy weather. Spray to deter pests if necessary.
- 'Tumbler' tomatoes are good for hanging baskets and boxes as they spill naturally over the sides and trail prettily. The tomatoes are also fairly small, sweet and juicy.
- Instead of cos lettuces, why not try 'Little Gem' or red-tinged 'Lollo Rosso' for a change.
- Ensure the rootball of the tomato plant is well below the compost level to prevent drying out. Water in well.

Strawberry Basket

A strawberry basket will bring delicious results and, with its white flowers and lustrous green foliage, will look attractive while you wait for it to bear fruit. All standard strawberry plants have pretty, white flowers; some have variegated foliage. Those with pink flowers look particularly attractive, yet some do not have sweet-tasting fruit. When choosing plants, ask your nursery or garden centre for advice.

Early planting, in mid- to late spring, will give you a first strawberry crop in early summer, with another later in the season. To avoid frosts, keep your basket in a greenhouse but do prop the door open on warm days to allow bees to enter and pollinate the flowers or else they will not produce fruit. Alternatively wrap your basket in frost-proof fleece at night during the early part of the year.

YOU WILL NEED

a wire hanging basket, 35 cm (14 in) in diameter · moss · slow-release fertiliser
5 strawberries 'Strawberries and Cream'

PERIOD OF INTEREST
May to July

LOCATION
In sun, hung in a greenhouse for the early months if possible, with the door open by day to allow pollination.

FOR CONTINUITY
Infill with alpine *Campanula carpatica* 'Blue Moonlight' or 'Isabel' to maintain colour interest.

1 Line your basket with moss, bringing it up and over the brim to hide the wire. Fill with rich compost until it is level with the top. Add the slow-release fertiliser after first checking that it is suitable for fruit.

2 Remove the plants from the pots and tease out their roots. Insert each plant at an angle all around the basket so that some will trail. If you can, plant some through the sides and base. Top with compost and water well.

GOOD ADVICE

- Feed your basket of strawberry plants regularly.
- Water when the compost is dry – but not too frequently as strawberries are particularly prone to mildew.
- Trim off any runners so that the plant's energy will then be used to produce larger and sweeter fruit.
- Look out for Perpetual (Remontant) varieties as these have a long fruiting season, beginning in early summer and then fruiting again in late summer. Other varieties may fruit only once a year.
- Guard against damage by squirrels and birds by covering with a fine-mesh plastic net while the fruit is forming.
- Pick the ripe strawberries every few days – do not remove them when still partially green or white – ripening produces natural sugars.
- Immediately after picking the fruit, cut off the leaves to about 7.5 cm (3 in) above the crown.

Summer Magic

Container gardening is addictive and the deeper you become immersed in the hobby the more success you'll enjoy. In no time at all you'll find you want to take on new challenges, and this usually means propagating your own plants from seed and cuttings. If you're a new recruit, this recipe of Junior petunias and dwarf coneflowers is easily achieved by buying in young plants in May and June. By late July you'll be revelling in a sea of blooms. Alternatively, if that's too easy or you want more for your money, both plants can be raised from seed sown in March in a heated propagator or on a windowsill indoors. Whichever way you choose, don't bury the plants too deeply in their display container nor overfirm the compost, or they will suffer through lack of air at the roots.

YOU WILL NEED

an oval or rectangular, terracotta trough, 40 cm (16 in) x 20 cm (8 in) · 3 dwarf coneflowers (*Rudbeckia hirta* 'Becky Mxd') 3 Junior petunias (*P.*, Fantasy Series, Milliflora Group)

PERIOD OF INTEREST
June to September

LOCATION
In full sun. Why not stand the trough on the edge of a border with other daisies planted behind it.

FOR CONTINUITY
Pinch off the spent petunia flowers, so they don't set seed. The coneflowers need only occasional deadheading.

1 Cover the drainage hole with broken crocks or, if these are difficult to obtain, a flat piece of polystyrene. Add more broken-up polystyrene in the base. This will save potting compost, particularly in a deep pot.

2 If heavy, move the pot into its display area before filling it with potting compost to within 2.5 cm (1 in) of the rim. Gently tease apart any plants (such as the coneflowers here) that have been grown in a seed tray.

3 Pick off any broken or yellow leaves. Plant the coneflowers, then add the petunias along the front. Top up the compost if necessary, then tap the trough once or twice to settle the compost. Finally, water well with a fine rose.

GOOD ADVICE

- If you are planting up a brand-new, terracotta pot, immerse it in water overnight to prevent the clay pot drawing out excess moisture from the potting compost.
- Add water-retaining granules to the potting compost if you want to cut down on watering.

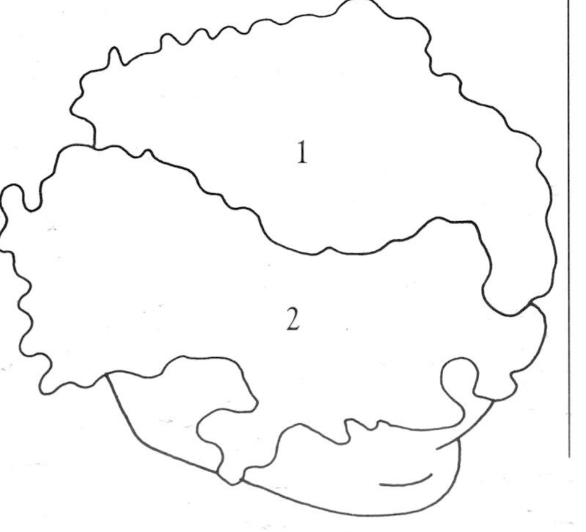

1 *coneflowers*
2 *petunias*

Summer Pastels

Soft colours will always be in demand when a delicate touch is called for in the garden, though it's not often that such colours are found in a packet of seeds sold as a mixture rather than separate individual colours. *Nemesia* 'Tapestry' (syn. *N.* 'Pastel') includes a scrumptious range of bicolours, pale yellows, golds, faded pinks and purples and is quite capable of making a feature on its own. It is simplicity itself to grow from seed sown in April and is quickly into flower. However, as bedding plants go, it is at its best for weeks rather than months, so I like to plant it behind a more permanent edging such as these grey-leaved senecios and brachycome daisies, which give some substance to the container when the nemesia has gone over. I thought the mellow, grey trough was the perfect foil for the flowers and its medieval carving linked in rather well with the general tapestry theme of the nemesia.

YOU WILL NEED

a concrete trough, 45 cm (18 in) x 23 cm (9 in) · 2 *Brachycome* 'Pink Mist'
3 *Senecio cineraria* · 24 *Nemesia* 'Tapestry'

PERIOD OF INTEREST
June and July

LOCATION
In full sun. Troughs look better on feet, raised up off the ground.

FOR CONTINUITY
Replace the nemesia with four pot-grown *Dianthus chinensis* 'Strawberry Parfait'.

1 Half fill the trough with potting compost. Space out and plant up the edging, alternating the brachycome daisies with the grey-leaved senecios. Angle the daisies slightly forward over the edge.

2 Split the nemesia into big chunks to drop in behind the edging plants. Such an unconventional transplanting method is far better than splitting a trayful of nemesia into individual plants.

3 Carefully set the nemesia chunks in position in the trough. Fill in any gaps with more potting compost. Then water the plants well to settle them in.

GOOD ADVICE

- Sow nemesia at intervals of 2–3 weeks to get a succession of blooms throughout summer.
- Overwinter the grey-leaved senecio in a sheltered spot outdoors or move it into an unheated greenhouse.

1 *brachycome*
2 *senecios*
3 *nemesia*

Gracious Herbaceous

In the early summer rush to select plants for containers, it is easy to have tunnel vision and consider just bedding plants. However, herbaceous plants can prove an attractive alternative. This gathering is full of character and has more variation in growth habit than a pot of, say, red salvias, French marigolds and busy lizzies. Herbaceous plants will give an almost instant effect when pot-grown, because many will have achieved their full height (if not spread) when you buy them. You can also enjoy them at close quarters in a pot while in season, and later plant them out permanently in a border to get bigger and better each year.

YOU WILL NEED

a terracotta pot, 45 cm (18 in) in diameter
1 red valerian (*Centranthus ruber* var. *coccineus*) · 2 *Salvia* x *sylvestris* 'Mainacht' ('May Night') · 1 *Heuchera micrantha* 'Palace Purple' · 2 cranesbill (*Geranium cinereum* 'Ballerina')

PERIOD OF INTEREST
May to August

LOCATION
In full sun on a patio. Add other pots around the base to build up the display.

FOR CONTINUITY
Dig out the salvia and valerian in July after flowering, add fresh compost and fill in with Japanese anemones (*Anemone* x *hybrida*) and *Rudbeckia fulgida* var *sullivantii* 'Goldsturm'.

1 Select your herbaceous plants from the wide variety available in pots in early summer. Mix flowers with coloured foliage for a balanced display.

2 Half fill the pot with potting compost and plant the taller plants – the valerian and salvia – in the back. Then put the heuchera in the centre and finally slip the cranesbill in the front so it can spill over the sides. Top up the compost and firm gently, then water the plants to settle them in.

GOOD ADVICE

- When you buy plants, look for a really full pot. If you plant up your pot in April, you may be able to split a single plant into two or even three separate pieces – a real bargain.

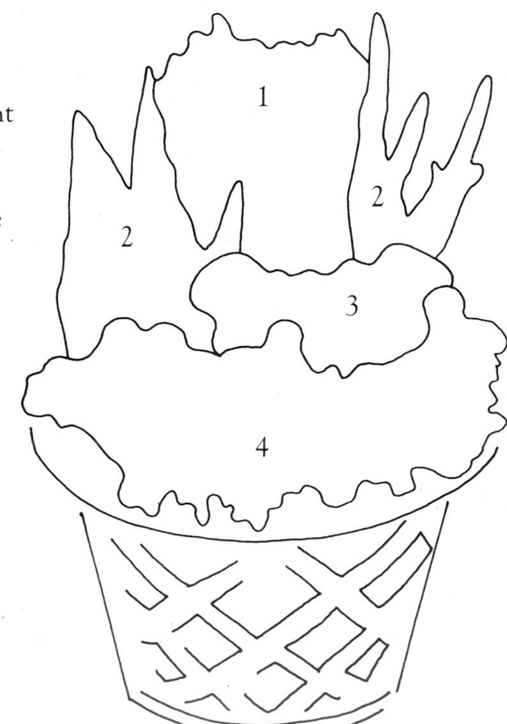

1 *valerian*
2 *salvia*
3 *heuchera*
4 *cranesbill*

Evening Shadows

Busy lizzies are certainly amongst the most obliging of summer bedding plants. They thrive in sun and shade and flower themselves silly without the need for deadheading or pinching out, but there are times when they can be a little – well how can I put this – boring. With this is mind, I cast a critical eye over John Betteley's plans for his handsome, simulated-stone urn and suggested that he consider adding a touch of drama in the form of some black pansies that I just happened to have in the back of the car. To set these off, we planted yellow-leaved creeping Jenny to drape over the front edge. By mid-summer the display had filled out beautifully. This project proved to be a useful exercise in choosing plants to suit certain conditions (in this case, semi-shade) as well as for their colour.

YOU WILL NEED

a simulated-stone urn, 40 cm (16 in) in diameter · 3 black pansies (*Viola* x *wittrockiana*) · 2 yellow-leaved creeping Jenny (*Lysimachia nummularia* 'Aurea') 1 diascia · 5 busy lizzies (*Impatiens*)

PERIOD OF INTEREST

June to September

LOCATION

In sun, shade or semi-shade. Lift your urn on to a pedestal for greatest impact.

FOR CONTINUITY

Deadhead the pansies twice a week. If they get straggly, cut them back hard and they will bush out again from the base.

1 Pot-grown pansies frequently have a tight mat of roots woven around the base and sides of the rootball. Although it seems drastic, pull off this mat to encourage the plant to grow away rapidly after replanting.

2 Tidy up the pansies by removing any yellow and spotted leaves, faded flowers and seed pods. Fill the urn with potting compost to within 5–7.5 cm (2–3 in) of the rim.

3 While still in their pots, arrange the display, placing the creeping Jenny at the front, then the pansies, backed by the diascia and busy lizzies. Plant them in the compost, topping it up if needed, and then water them in.

GOOD ADVICE

- Give a shallow urn like this plenty of water and fertiliser if it is planted as densely as this one.
- Position your urn where it can be seen and enjoyed from the house windows.

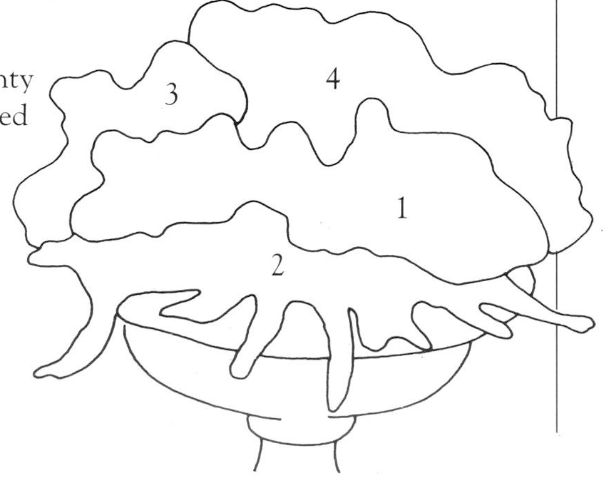

1 *pansies*
2 *creeping Jenny*
3 *diascia*
4 *busy lizzies*

Hosta Power

If I were to list my top ten plants for container gardening, hostas – or plantain lilies – would be one of the first I'd think of. They have impeccable credentials: as well as having handsome foliage (and spikes of trumpet-shaped flowers), they will grow for years in a decent-sized pot in sun or shade. Their main drawback is that they attract slugs and snails, though growing hostas in a pot significantly reduces damage. Pick off any culprits in the evening as they emerge from each hosta's crown.

With such a symmetrically shaped plant, the usual approach is to give each variety a pot of its own and group the pots together. However, mixing and matching plants is half the fun, so I picked out two of the most spectacular and diverse hosta types and contrasted their habit with a trailing campanula.

YOU WILL NEED

a terracotta pot, 50 cm (20 in) in diameter
1 blue-leaved hosta (*H. sieboldiana* var. *elegans*) · 2 variegated hostas (*H.* 'Ground Master') · 1 *Campanula poscharskyana*

PERIOD OF INTEREST
May to October

LOCATION
In sun or shade. Sit on paving around a pond or other water feature.

FOR CONTINUITY
Plant spring bulbs like dwarf daffodils to flower before the hosta leaves emerge. Leave the old, dried flower spikes on in winter. They look quite sculptural.

Pick out hostas with several shoots growing up in the pot so you can get a really good-sized spread of leaves. Almost fill the pot with the potting compost. Plant the blue-leaved hosta at the back as it's the most vigorous. Then add the variegated hostas and campanula and water well.

GOOD ADVICE

- Pull away old campanula stems after the flowers have faded.
- Some gardeners smear grease around the pot rim to deter slugs and snails.
- Some thinner-leaved, variegated hostas can scorch in intense sunlight or shrivel if too dry, so site them in shade.

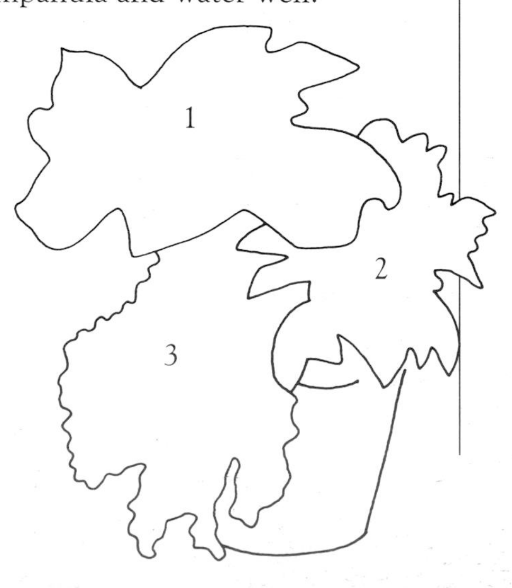

1 *blue-leaved hosta*
2 *variegated hosta*
3 *campanula*

Dapper Dahlias

Have you noticed that dahlias are rapidly shrinking? This is not such good news if you're after a great armful of blooms for harvest festival, but is great for adventurous container gardeners who want to enjoy these sparkling extroverts en masse and in a confined space. Long, narrow, terracotta troughs, as used here, do not hold that much potting compost, so you need to choose your plants carefully: African marigolds, for example, would quickly grow top-heavy and starve in such a container. My choice of dahlias here was tailored to form a back row, with the fluffy, powder-puff flowers of ageratum in front, bordered by bush lobelia. The lobelia helps to soothe and smooth over any potential colour clashes from the dahlias behind, which is just as well because they are sold in mixed colours and you can't always see what's coming!

YOU WILL NEED

a terracotta trough, 60 cm (24in) x 20 cm (8 in) · 4 dahliettas or other similar dwarf varieties that do not exceed 30 cm (12 in) · 2 floss flowers (*Ageratum*) · 3 bush lobelia (*L. erinus* 'Crystal Palace')

PERIOD OF INTEREST
June to September

LOCATION
In sun, on a recessed windowsill or low wall or on a balcony or patio.

FOR CONTINUITY
Deadhead the dahliettas twice a week.

1 Either grow your plants from seed sown in February or, as I did, buy them from a nursery or garden centre in late May or early June. Check heights and colour range on the label before you buy.

2 Fill your trough three-quarters full with potting compost. Space out the dahliettas evenly along the back and plant them. Then plant the floss flowers and lobelia, alternating them at the front of the container.

3 Tap down the trough a couple of times to settle the compost, before watering in the new plants. Then sit back and look forward to a bumper display!

GOOD ADVICE

- Dahliettas tend to flower in flushes rather than continuously, so keep them well watered and fed to encourage the next instalment.

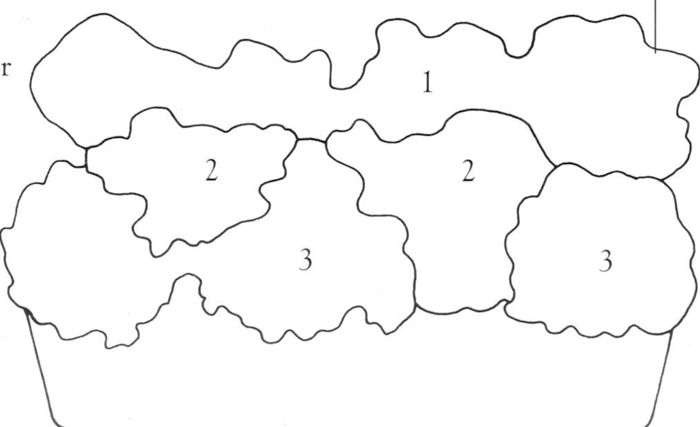

1 *dahliettas*
2 *floss flowers*
3 *lobelia*

Dynamic Duo

Virtually every plant has a perfect partner and, when you've found it, you get a great sense of achievement and an urge to share the idea. Well my dynamic duo is cosmos and petunias. The choice of cosmos is however vital: most would be far too tall and quickly shade out an underplanting so you have to stick to dwarf varieties, and 'Sonata Mxd' is the best I've grown. It is so eye-catching that it's being increasingly sold as a pot-grown bedding plant alongside more familiar favourites like begonias, lobelia and ageratum. However, it's easy to raise from seed and you can have ten times as many plants for a fraction of the price. Raising petunias from seed requires more heat and they're slower growing, so – as I did – you might decide to buy these, in which case pick out any pinks and purples that match the cosmos.

YOU WILL NEED

a wicker basket, 50 cm (20 in) in diameter
polypropylene or plastic liner
polystyrene chunks · 5 cosmos
(*C. bipinnatus* 'Sonata Mxd') · 5 petunias
(*P.* 'Celebrity Ovation') · hay

PERIOD OF INTEREST
June to September

LOCATION
In a sunny spot. A background of greenery will highlight the blooms.

FOR CONTINUITY
Cut off spent cosmos flowers before they set seed and pinch off fading petunias. Liquid feeding will prolong the show.

1 Sow your cosmos seed in April in a greenhouse or indoors on a windowsill. You don't need artificial heat. After two weeks prick out into cell trays, three seedlings to a cell.

2 When the cosmos plants are about 15 cm (6 in) tall, transfer them to larger pots or straight into your display basket (see below) to grow on.

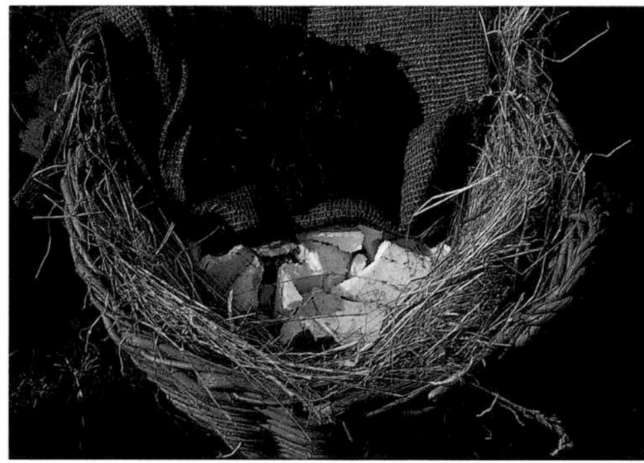

3 Plant up your basket in early June. Line it first, pack out the base with the polystyrene chunks, then add the potting compost and plants. To create a country flavour, tuck in an edging of hay. Then water well.

GOOD ADVICE

- When buying petunias, look for bushy plants with dark green leaves.
- Petunias dislike heavy rain so site your basket where it will get some protection under a house wall or an open veranda.

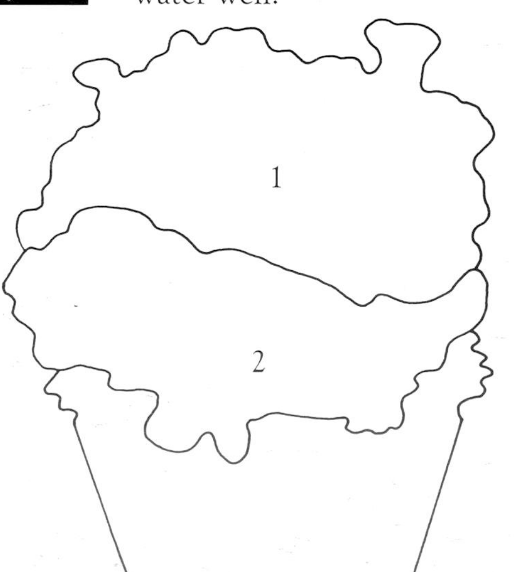

1 *cosmos*
2 *petunias*

Cottage Pot

To me, the cottage-garden style is typified by a profusion of modest flowers growing together in one great tumbling mass without any conscious attempt at colour theming, separating roses from hardy perennials, vegetables and fruit, or keeping plants rigidly within bounds. I have here recreated just such a look by using hardy perennials (lady's mantle and cottage pinks), a tender perennial (osteospermum), a hardy annual (cornflower) and a half-hardy annual (floss flower). By starting with a suitable 'weaver' at the front of the pot – like this greeny yellow lady's mantle – you can end up with a glorious sea of intermingled colour as the 'weaver' threads itself through the other plants and then comes into flower.

YOU WILL NEED

a terracotta pot, 40 cm (16 in) in diameter
1 lady's mantle (*Alchemilla mollis*)
6 dwarf blue cornflowers (*Centaurea cyanus* 'Jubilee Gem') · 1 purple osteospermum (O. 'Sunny Lady') · 5 red cottage pinks (*Dianthus* x *allwoodii* 'Cheryl')
2 purple floss flowers (*Ageratum*)

PERIOD OF INTEREST
June to August

LOCATION
A sunny, sheltered spot will encourage flower production.

FOR CONTINUITY
Strip out the pot, retaining the floss flowers and osteospermum, top up with fresh compost and plant pot-grown asters or coneflowers.

1 Dig up a crown of lady's mantle from yours or a friend's border in March and pot it up, or buy one in May. Grow the cornflowers from seed in April; they are the only ingredients that you're unlikely to be able to buy.

2 Assemble the plants and water them well. Fill the pot with potting compost to within 5 cm (2 in) of the rim.

3 Plant the lady's mantle towards the front, backed by the osteospermum. Work the cornflowers and cottage pinks into the gaps. Add the floss flowers. Water well.

GOOD ADVICE

- For a range of cornflower colours, try *Centaurea cyanus* 'Polka Dot Mxd'.
- Spend some time 'dressing' the pot as it matures, spacing out the blooms almost as if it were a flower arrangement.

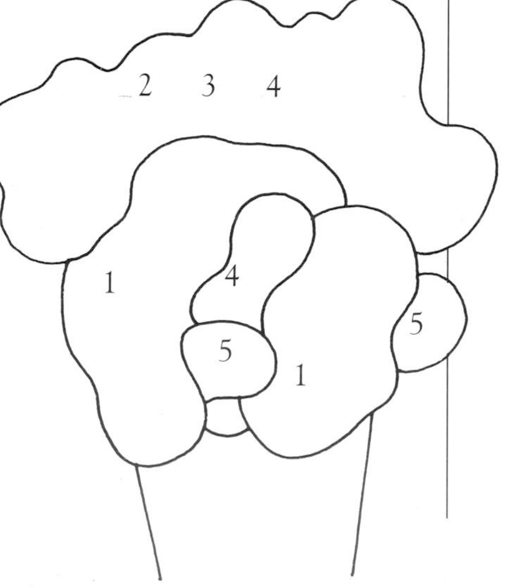

1 *lady's mantle*
2 *cornflowers*
3 *osteospermum*
4 *cottage pinks*
5 *floss flowers*

A Class Act

I'm not a plant snob, but I do like to gather around me a handful of the most refined and sophisticated plants in the same way that wine buffs stock their cellar with the finest vintages. At the top of my list must come tuberous lilies for they are beguilingly beautiful and – as luck would have it – they perform very well in containers of all shapes and sizes, even small ones like this terracotta trough. The secret to a successful display is to match the vigour of your lilies to the size of the container and its companions; then think up a subtle (or dazzling) colour scheme and you're almost there. The last factor to consider is whether the plants will be in sun or shade. My combination of lilies, an exquisite fern and double busy lizzies will be happier in shade or semi-shade than full sun, and they'll last longer too.

YOU WILL NEED

a decorated, terracotta trough, 50 cm (20 in) x 25 cm (10 in) · 5 red or pink dwarf lilies (*Lilium* 'Mr Edd') · 2 double red busy lizzies (*Impatiens*) · 1 painted fern (*Athyrium niponicum* var. *pictum*)

PERIOD OF INTEREST
June to September

LOCATION
In semi-shade near the house or a seat where you can enjoy the lilies' delicious scent.

FOR CONTINUITY
Replace the lilies after they have flowered with three pot-grown fuchsias.

1 Lilies in garden centres are usually sold just coming into bloom, but if you want to be sure of a specific variety and save money, pot up your own tubers in March and April and grow them on outdoors under a warm wall.

2 Gather your plants together. You can buy larger busy lizzies than these, and well in bloom, but these are typical of starter plants bought in April, potted up and grown on for 3–4 weeks.

3 Fill the trough three-quarters full with potting compost. Plant the fern first, handling its brittle fronds with care. Add the lilies along the back, then the busy lizzies. Top up the compost and water well using a fine rose.

GOOD ADVICE

- After flowering, plant out the lilies in a border with their roots in shade and their heads in sun.
- For a succession of lily blooms, pot up the tubers every 2–3 weeks.
- Busy lizzies are very tender so don't plant them out until early to mid-June.

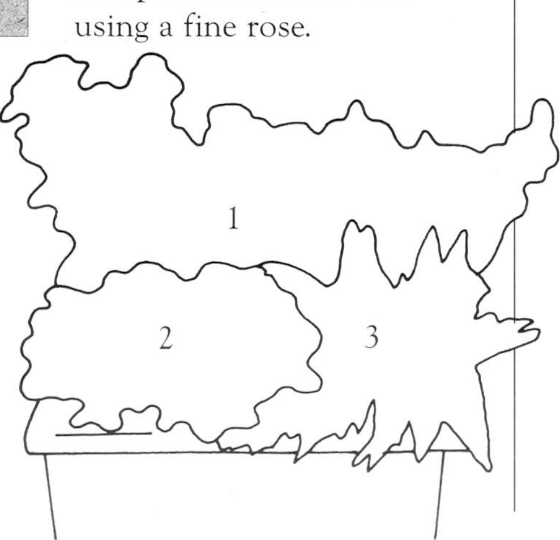

1 *lilies*
2 *busy lizzies*
3 *painted fern*

Blazeaway Begonias

'Non Stop' begonias are aptly named. They never give up for a moment though their flowers don't get anywhere near the size of those dinner-plate blooms you see at flower shows, which need crutches and splints to hold them up. 'Non Stop' begonias are highly sociable and can be mixed with other summer flowers like busy lizzies, fuchsias and trailing lobelia that share a love of shade (though they will also perform well in full sun if well watered).

For a really eye-popping show, why not give these begonias a container to call their own. Being slightly pendulous, they will spill forwards and help to soften the edge of the tub, which, if it is decorated on the front like this one, need not be hidden beneath a cascade of trailers. As with busy lizzies, it's well worth picking off spent blooms and petals to keep the plants shipshape.

YOU WILL NEED

a reconstituted-stone planter, 38 cm (15 in) square · 5 *Begonia* 'Non Stop' in mixed colours

PERIOD OF INTEREST
June to September

LOCATION
In sun or shade. Plant up more 'Non Stop' begonias in a bed adjacent to your pot to create a cascade and river of colour.

FOR CONTINUITY
Don't let the begonias dry out and apply a liquid feed every 7–10 days.

1 Pot up your begonia corms in spring, setting them just below the compost surface, hollow side up. Grow on indoors on a windowsill or in a heated greenhouse. At first, water sparingly and not directly on to the corm.

2 Once there is little danger of frosts (usually by early June), gradually harden off the plants by standing them outside against a warm wall; return them indoors if cold nights are forecast. Then almost fill the planter with potting compost and transplant the begonias, placing one plant in the centre and one at each corner. Top up the potting compost and water well.

GOOD ADVICE

- You don't need to deadhead these begonias, but it's worth picking up fallen flowers as they can mark the leaves while they decompose.
- Why not plant a selection of 'Non Stop' begonias, including the gorgeous tangerine included here, giving each colour a container to itself.
- Be adventurous and include some varieties grown for their beautiful leaves. *Begonias rex*, for example, will enjoy a summer holiday in the garden, given shade and shelter.
- You can keep your begonias from year to year by saving the corms. When the leaves have been touched by frost, dig them up and dry them off in a shed or greenhouse. Once the leaves and stems have fallen off, plunge them in old, dry potting compost for the winter, keeping them cool but frost-free. Start them off into growth again in spring by potting them up.

Everlasting Pleasure

Although perhaps more associated with dried indoor arrangements than border or container displays, the hot colours of strawflowers can transform a bare patch into a fiery furnace of colour and 'Bright Bikinis' is the best dwarf strain I've grown. They mix well with other sun-loving daisies like zinnias, but I chose to pick up the dry, bristly texture of the petals by packing them into a wicker basket with their pots buried in hay and edged up with a collection of dried sunflower heads that were allowed to wither *in situ* on their thick stems. I was so pleased with the results that next year I plan to add other 'dried flowers' such as statice and love-lies-bleeding in separate baskets to build up the display.

YOU WILL NEED

a wicker basket, 40 cm (16 in) in diameter and at least 15 cm (6 in) deep · 24 dwarf strawflowers (*Helichrysum bracteatum* 'Bright Bikinis') · 12 dried sunflower heads (*Helianthus*)

PERIOD OF INTEREST
July to September

LOCATION
In an open sunny spot. If the basket is strong enough it can even be suspended outside a window.

FOR CONTINUITY
Grow some extra strawflowers to plant out in a nursery bed for cutting and drying and combine them with the sunflower heads to repeat the theme indoors over the winter.

1 Strawflowers can be raised from seed sown indoors in April and pricked out into a seed tray. In 4–6 weeks, transfer the young plants into 12.5 cm (5 in) pots. Sunflowers can be sown straight into the border in May.

2 When the strawflowers are beginning to show colour, bury them – pot and all – in a mulch of hay within the wicker basket. Finish off the arrangement by inserting the dried sunflower heads.

GOOD ADVICE

- Snip off the strawflowers when they are fully expanded and are beginning to go brown in the centre to encourage more flowers to develop.
- For indoor decoration it is vital to pick strawflowers just as they are beginning to open.

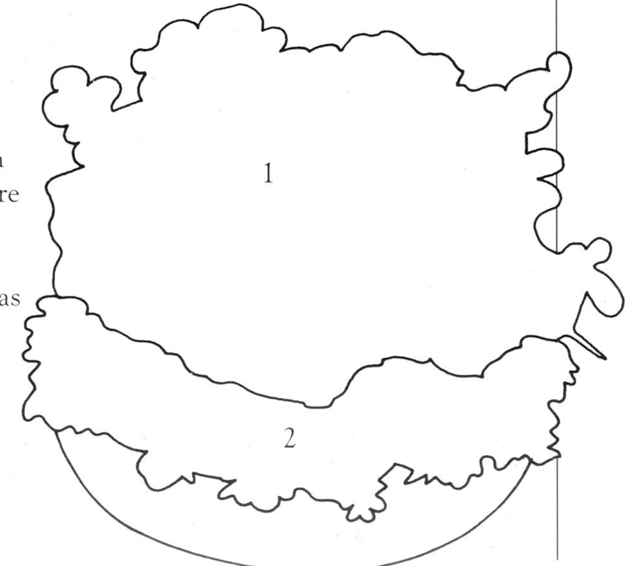

1 *strawflowers*
2 *sunflowers*

Flower Festival

If spring comes early and you haven't planted many late spring-flowering bulbs like tulips, you may find a bit of a gap developing in your container display in May, yet it is too early to plant out the summer bedding – except in only the very mildest parts of the country. For times like this, varieties that will tolerate a few cold nights or even a touch of frost are a godsend, and osteospermums – those prolific South African daisies – are amongst the best. Pansies, too, are good for it is now that they peak. There's a good chance you will have pansies in full bloom, but if not there are still plenty on sale. With luck, you may even find the lovely, ginger-coloured flowers that I've used in this flower festival display, and which harmonise so well with the terracotta.

YOU WILL NEED

a shallow, terracotta bowl, 50 cm (20 in) in diameter · 3 contrasting osteospermums (O. 'Sunny Lady', O. 'Buttermilk' and O. 'Silver Sparkler') · 5 pansies (*Viola* x *wittrockiana*) in mixed colours
broken crocks

PERIOD OF INTEREST
May to September

LOCATION
In full sun. For greater impact, raise the bowl up higher on an upturned pot.

FOR CONTINUITY
By July the pansies will begin to get leggy so cut them back hard and plant them in a border. Replace with pot-grown coneflowers.

1 Select your osteospermums and pansies in May. There are usually plenty of variations to choose from. Water them thoroughly. Then fill the bowl to within 5–7.5 cm (2–3 in) of the rim with potting compost.

2 Put the three osteospermums in position, spaced around the bowl, then fill in around them with the pansies. Top up with potting compost. As a finishing touch, use some broken crocks to make an edging around the bowl. Water well to settle in the plants.

GOOD ADVICE

- Broken crocks placed around the edge of the bowl not only look good but also stop the potting compost washing over the edge.
- Regular deadheading is essential to keep up the show.
- Overwinter the osteospermums in cool, frost-free conditions.

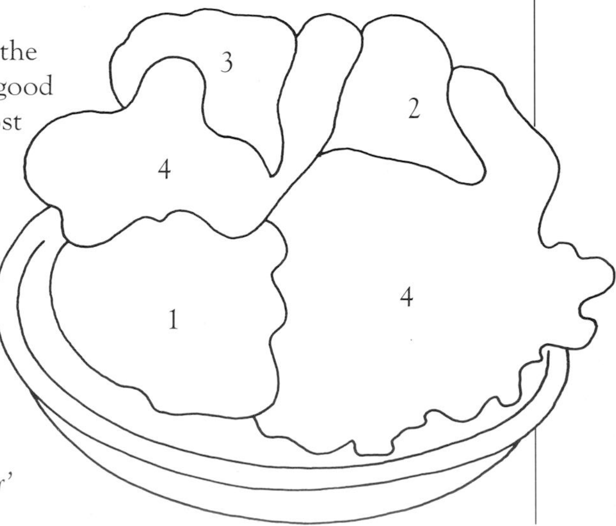

1 Osteospermum *'Sunny Lady'*
2 Osteospermum *'Buttermilk'*
3 Osteospermum *'Silver Sparkler'*
4 *pansies*

In the Pink

Searching for a theme can often cause problems for the container gardener faced with a wide range of (empty) containers and an almost overwhelming choice of plants. So, in the same way as when I'm designing a garden, I look for a suitable starting point. In this case it was some silver-leaved senecios left over from the previous year. They were a bit unkempt but redeemable and would provide a perfect foil for a pastel scheme.

Pink seemed the obvious choice. It is the most popular, bedding plant colour, though it can clash with new, orangey red terracotta. Luckily my pot was well weathered with moss and chalk deposits, which would echo the grey and green in the planting. I then chose a tall plant, two bushy plants and three trailers to create a well-balanced display.

YOU WILL NEED

a well-weathered, terracotta pot, 45 cm (18 in) in diameter · 3 *Senecio cineraria* 'Cirrus' · 1 pink marguerite (*Argyranthemum* 'Vancouver') 2 *Pelargonium* 'Lovesong' · 3 *Verbena* 'Sissinghurst'

PERIOD OF INTEREST
June to September

LOCATION
In full sun, with plenty of space around the pot so the verbena can spread.

FOR CONTINUITY
Deadhead the pelargonium and marguerite every week. Cut off any flowering shoots on the senecio.

1 Assemble the plants, and water them thoroughly until they are all nicely moist in their containers.

2 If you are refurbishing a pot that already contains some plants, carefully dig around them with a trowel; try not to disturb their roots unduly. Then remove the top 15 cm (6 in) of potting compost.

3 Top up with fresh potting compost to within 7.5 cm (3 in) of the rim. Plant the marguerite towards the centre, flanked by the pelargoniums and then put the trailing verbenas around the outside. Water the pot well.

GOOD ADVICE

- You can speed up the ageing process on a new terracotta pot by painting on live yoghurt.
- Silver-leaved senecios can easily be raised from seed or bought as young plants.

1 *senecio*
2 *marguerite*
3 *pelargonium*
4 *verbena*

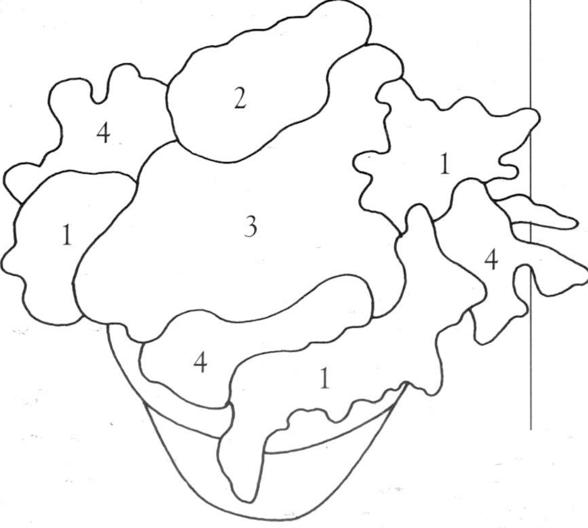

Trail Blazers

One of the most exciting plant introductions of recent years has been a new race of super-vigorous hybrid petunias that can put on up to 2 m (6½ ft) of growth in a single season. They now sell by the million as young, pot-grown plants or smaller plug plants and are all struck from cuttings taken from mother plants at the nursery. If you are not used to such rampant growth rates, these hybrid petunias can quickly get the better of you, so my advice is threefold.

Firstly, select a tall container so that the petunias can cascade over the edge. Secondly, pair up these petunias only with equally robust plants, and, thirdly, keep pinching them out, especially in the early stages of growth, so that you encourage a bushy cascade, not just a few overlong trails.

YOU WILL NEED

a reconstituted-stone urn, 35 cm (14 in) in diameter, sitting on a pedestal
polystyrene chunks · 2 *Petunia* 'Surfinia'
4 tapien verbenas (*V.* 'Tapien Pink' and *V.* 'Tapien Violet') · 2 *Helichrysum petiolare*

PERIOD OF INTEREST
June to September

LOCATION
In sun, raised up in a border and backed by planting or against a wall or hedge.

FOR CONTINUITY
Liquid feed every 7–10 days or add a slow-release feed, which works for a whole season, to the potting compost.

1 Pack out the bottom third of the urn with broken polystyrene and then fill up with potting compost to within 5–7.5 cm (2–3 in) of the rim. Assemble your plants and water well.

2 Move the pots around until an attractive display is achieved. Then plant the petunias in the middle, and space out the verbenas around them. Put the grey helichrysums on opposite sides.

3 Finally pinch out the petunias to build up a strong framework of shoots, and water the compost thoroughly to settle in the plants.

GOOD ADVICE

- Pinch out the grey helichrysums a few times to encourage them to bush out.
- If you are prepared to wait a bit longer for the urn display to fill out, reduce the plant quantities by half.

1 *petunias*
2 *verbenas*
3 *helichrysums*

Ice and Lemon

All-white colour schemes (if white is a colour) have never really gone out of fashion, but on dull, overcast days they can look a little dreary. One solution is to add a splash of yellow, which gives white leaves and flowers a really refreshing zing. In early summer there is plenty of yellow to pick from, though the two in my ice-and-lemon cube are amongst the most outstanding and richly coloured; they will also flower continuously till the frosts arrive.

My favourite container creations are ones that feature foliage and flowers, and ornamental cabbage and kale are worthy contenders. As the evenings get cooler, the leaf colours intensify and peak in late summer and autumn. As well as white, there are pink and purple selections to fire your imagination.

YOU WILL NEED

a terracotta pot, 40 cm (16 in) square
1 ornamental cabbage Northern Lights Series · 2 *Calceolaria* 'Sunshine'
1 strawflower (*Helichrysum bracteatum* 'Dargan Hill Monarch') · 2 white petunias
2 white New Guinea busy lizzies (*Impatiens*, New Guinea Hybrids)

PERIOD OF INTEREST
June to September

LOCATION
In sun or semi-shade. A pair would look good flanking a door.

FOR CONTINUITY
Pinch out fading petunia flowers before they run to seed and snip off the oldest strawflowers as they discolour.

1 After assembling your plants, dunk any that are dry in a bowl of water for a couple of hours before planting. (Ornamental cabbages, calceolarias and strawflowers can be bought as younger plants if you prefer.)

2 Half fill your container with potting compost. Then, with the plants still in their pots, make a pleasing arrangement, keeping the taller plants – cabbage, calceolarias and petunias – to the rear of the planter.

3 Working from the centre to the edge of the planter, carefully knock each plant out of its pot – protecting its roots and the rest of the plants from its compost – and plant it up. Top up the potting compost and water well.

GOOD ADVICE

- To add an 'antique' finish to a new terracotta pot, paint on a thin solution of lime and water, as I've done here. Always wear protective gloves and goggles.

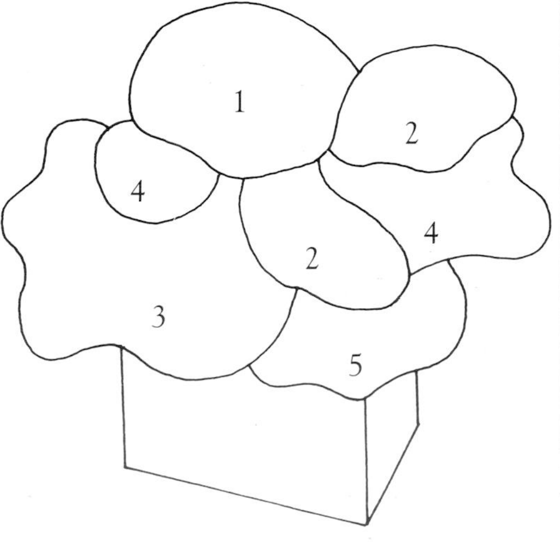

1 *ornamental cabbage*
2 *calceolarias*
3 *strawflowers*
4 *petunias*
5 *busy lizzies*

Monkey Business

Monkey musk, or mimulus, is amongst the most extrovert of garden flowers, and it reveals an almost-human quality in the way its blooms peer around to face each other. Its particularly vivid flowers are often speckled and spotted within, as in a deep-throated orchid. Monkey musks love moisture and the more they get the merrier they will be. This fact more than compensates for their somewhat restricted flowering period compared to, say, the trailing lobelia that I have partnered them with here. If you are brave enough, let the lobelia wander through the bright yellow leaves of the deadnettle and yellow-leaved creeping Jenny, and every tiny flower will be thrown into sharp focus. Fortune favours the brave as they say.

YOU WILL NEED

a terracotta pot, 40 cm (16 in) in diameter
2 plain monkey musk (*Mimulus* 'Malibu Orange') · 1 spotted monkey musk (*Mimulus*) · 2 yellow-leaved deadnettles (*Lamium maculatum* 'Aureum') · 3 yellow-leaved creeping Jenny (*Lysimachia nummularia* 'Aurea') · 2 trailing lobelia (*L. richardsonii*)

PERIOD OF INTEREST
June and July

LOCATION
In half sun or full shade. Hostas would look good alongside.

FOR CONTINUITY
Replace the monkey musk with pot-grown dwarf coneflowers or chrysanthemums.

1 Sow the monkey musk in March in a heated greenhouse or on an indoor windowsill. Or, buy plants as they are coming into bloom along with the lobelia and foliage plants. Almost fill the pot with potting compost.

2 Interplant the deadnettle and creeping Jenny along the front edge, then add the two lobelias behind. Finally plant the monkey musks at the back. Top up with more potting compost if necessary, and then give the plants a good soaking to settle them.

GOOD ADVICE

- Overwinter the perennial *Lobelia richardsonii* in frost-free conditions and bring it into active growth again the following spring.

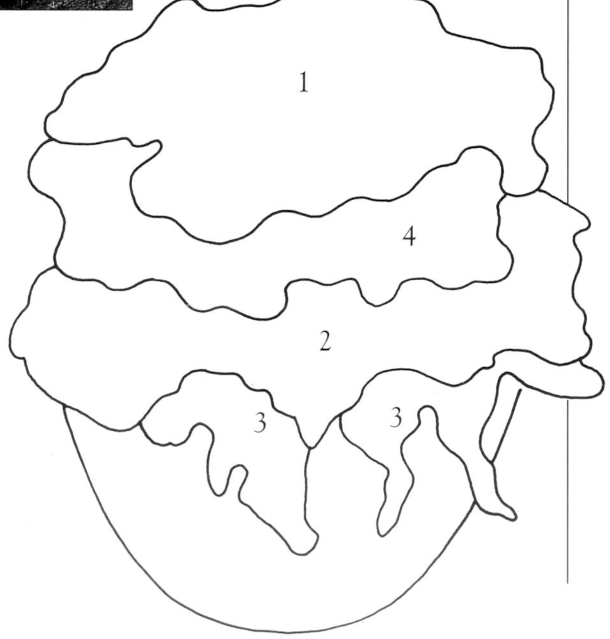

1 *monkey musk*
2 *deadnettle*
3 *creeping Jenny*
4 *lobelia*

Salad Days

Digging up parsnips and carrots has a certain excitement to it, but it can't match the pleasurable anticipation when the first, home-grown salad crops are imminent. Even if tomatoes were inedible (they were at first thought to be poisonous), the dwarf bush and trailing varieties would still be worth growing just for their ornamental value. 'Red Robin' – an ultra dwarf tomato and less domineering than the more familiar 'Tumbler' – makes an excellent focal point in an edible container.

For a longer-lasting show, select varieties you can pick over like 'Lollo Rossa' lettuce rather than those that are cut whole, leaving a gap. You can also mix in some of the more refined culinary herbs like basil and parsley, while tagetes will inject some extra colour and (some believe) deter sap-sucking insects.

YOU WILL NEED

a plastic pot, 40 cm (16 in) square
2 dwarf bush tomatoes 'Red Robin'
3 pale orange tagetes (*T.* 'Tangerine Gem')
2 parsley (*Petroselinum*) · 9 basil (*Ocimum basilicum*) · 3 dwarf lettuce 'Little Leprechaun'

PERIOD OF INTEREST
June to September

LOCATION
In sun and shelter. Near the kitchen door would be convenient.

FOR CONTINUITY
Sow two or three batches of lettuce in small pots to replace harvested plants.

1 Select your favourite salad vegetables and herbs. They can be grown from seed or bought as young plants. Sow the basil seed indoors in May; basil is very sensitive to cold so don't sow it any earlier than this.

2 Almost fill the pot with potting compost. Plant the tomatoes towards the back surrounded by the tagetes and flanked by the parsley. Add a row of basil, then dwarf lettuce to edge up.

3 Top up the compost to a maximum of 2.5cm (1in) below the rim, so the compost does not get washed over the edge while watering. Place the pot under glass if you are keen for quick results.

GOOD ADVICE

- Pick over the herbs regularly to keep them bushy and compact.
- Feed all the plants every week with a tomato fertiliser.

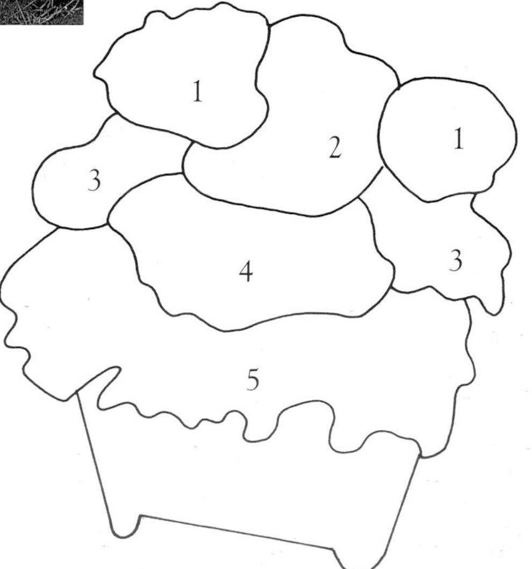

1 *tomatoes*
2 *tagetes*
3 *parsley*
4 *basil*
5 *lettuce*

Strawberry Surprise

Strawberries – as you doubtless know – come in all shapes and sizes, from the tiny delectable alpine forms to juicy monsters as big as a child's fist, but have you seen pink-flowered and variegated plants? They really are attractive, especially at flowering time, though I should warn you not to expect great bowlfuls of fruit. I counted eight strawberries on these three plants: the birds ate three and the children had the rest so I can't vouch for the flavour! Anyway, this display provides a real conversation piece that improves with age as the runners descend over the sides like mountaineers on ropes. Because the fruits are raised off the ground, they are more likely to escape attack by slugs and won't get splashed with mud. In fact, I've convinced myself that next year my strawberry surprise could well form the centrepiece of a mobile strawberry patch.

YOU WILL NEED

a clay pan, 30 cm (12 in) in diameter and 15 cm (6 in) deep · 2 pink strawberries 'Ruby Surprise' · 1 variegated strawberry fine stone chippings

PERIOD OF INTEREST
May to September

LOCATION
In sun, sheltered from cold winds.

FOR CONTINUITY
Add trailing lobelia between the strawberry plants to give colour between the flushes of strawberry flowers.

1 Assemble your strawberry plants in April or May. These specimen-sized plants will give almost instant maturity, but smaller ones will do. Put a layer of potting compost in the bottom of your pan.

2 Plant the pink and variegated plants. Fill in the gaps with more potting compost, ensuring you don't bury the centre of your plants with compost or they may rot. Use a sprayer to clean any compost off the leaves.

3 Finally add the finishing touch – a dressing of fine stone chippings. This prevents the potting compost from washing away and provides good surface drainage. Give the plants a good soaking to settle them in.

GOOD ADVICE

- As runners form, stand the pan up on another pot so they hang down freely, or peg them into small pots alongside and let them root.
- Watch out for infestations of greenfly. Squeeze them between your fingers to control them.
- As an alternative to 'Ruby Surprise', try 'Pink Panda'.

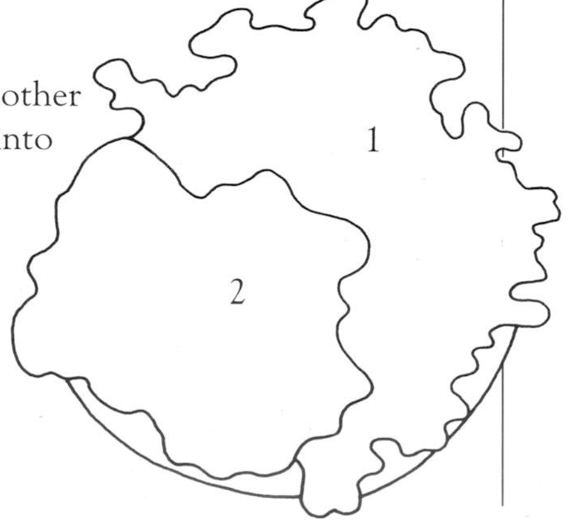

1 *pink strawberries*
2 *variegated strawberry*

Autumn Delicacy

All too often, as mid-autumn approaches and our bright summer baskets die off, gardeners empty their hanging baskets and windowboxes and put them away until spring.

The autumn and winter months, however, are just the time when your house and garden need extra life and colour. This can be very easily achieved by creating an arrangement of hardier plants that can withstand colder, less sunny conditions and still give a good display of gentle colour and texture.

Instead of the usual gold and yellow colours of autumn which, incidentally, can be achieved with a mix of miniature chrysanthemums, we have opted for a more delicate, silver-grey and pink scheme. Planted in mid-autumn, it will still be looking healthy in early spring.

YOU WILL NEED

a reconstituted-stone trough, 60 cm (24 in) x 23 cm (9 in) · 1 purple hebe (*H.* 'La Séduisante') · 1 white hebe (*H.* 'Kirkii') 1 scented wormwood (*Artemisia* 'Powis Castle') · 3 pink/white winter heathers (*Erica*) · 2 pink ornamental cabbages

PERIOD OF INTEREST
October to March

LOCATION
In sun, on a windowsill if possible; will tolerate semi-shade.

FOR CONTINUITY
Fill any gaps with delicately toning cyclamen providing the trough is not sited in an exposed position.

1 After ensuring that there are drainage holes, line the base of the box with crocks or pebbles. Half fill the trough with potting compost. Test your planting scheme, aiming to have the larger plants to the rear.

2 Once you are happy with your plan, firm the largest plants in first. Then infill with the heathers and ornamental cabbage along the front. Top up with compost, firm in gently and then water the arrangement.

GOOD ADVICE

- Use hebes only if you have a fairly tall window as some can grow to 2 m (6 ft)!
- Windowboxes can be used to give you extra privacy if the planting arrangement is tall.
- Heathers like a rich moist potting compost so keep the arrangement well-watered.

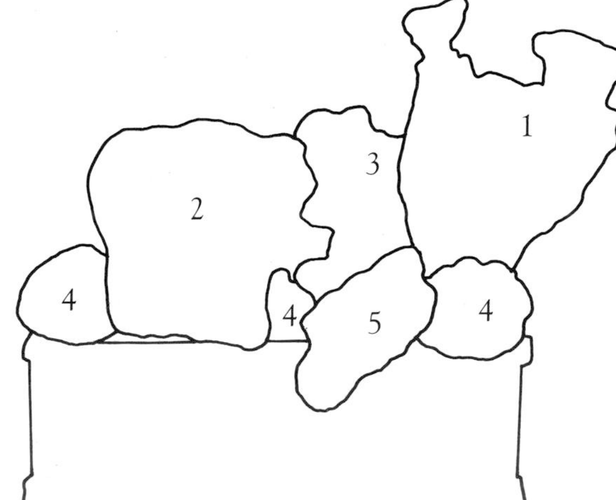

1 *purple hebe*
2 *white hebe*
3 *wormwood*
4 *heathers*
5 *ornamental cabbages*

Pansy Ball

One of the most reliable flowers for autumn and winter baskets has to be the pansy, but rather than the usual half hanging basket you could create a ball.

To do this, it is possible to wire together two standard half-baskets or you can now buy basket balls that clip together. If making your own, try to choose smaller baskets as, once filled with soil and water, they are very heavy.

In mid-spring, you can buy a tray of plug plants, which are easy to insert from the outside. If using a wire basket, you will need to line it with moss and polythene. To make planting easier, balance each half on a filled flowerpot.

Allow the baskets to settle for a few days before joining, or overfill one half. Otherwise there will be insufficient compost and moisture around the plants in the top section of the ball.

YOU WILL NEED

2 plastic baskets, each 35 cm (14 in) in diameter · polythene liner
48 pansies (*Viola* x *wittrockiana* 'Padparadja Forerunner Tangerine')

PERIOD OF INTEREST

October to May

LOCATION

In sun or semi-shade, in a doorway, where a splash of colour is especially welcome throughout winter.

FOR CONTINUITY

The ball could be reused in summer for felicia daisies or a ground-cover rose such as *R.* 'Hertfordshire'.

1 Insert the liner. Make holes at regular intervals around the base of the first half. Insert the plugs from the outside, roots first, protecting the roots in a polythene cone. Add layers of compost as you work upwards.

2 Overfill both baskets with compost. Water them well to settle in the compost. Then cover one basket with plastic, invert it over the second basket and slide the plastic out. Clip or wire the two baskets together.

GOOD ADVICE

- Overfill both baskets so that, when placed together, the compost is compact and there are no air pockets where plant roots could dry out.
- Wrap plant roots in a small cone of polythene to prevent damage. Polythene will slide through more easily than damp newspaper. Then remove.
- A small empty flowerpot buried in the top basket will make watering easier. If this pot will be visible when the ball is hung up, you could drop in a potted-up pansy and then simply remove it for watering.
- A plastic, clip-together ball can be planted later in the season, using full-sized pansies.
- To make gaps large enough to insert mature pansies from outside the ball, clip away the vertical or horizonal plastic struts if necessary.
- To maintain the strength of the ball, ensure that these enlarged gaps are not made directly beside each other.
- Remove chains while infilling with pansies but remember their positions while planting.

Everlasting Evergreens

Although many of us opt for evergreens such as ivy somewhere in our spring and summer baskets, it is rare to see an evergreen basket bringing life to a doorway in autumn and winter. However, bright-berried plants and tiny-flowered evergreens, with their glossy, red- or gold-tinged foliage, will last through some of the coldest weather and can bring a great deal of pleasure.

Most evergreens are far hardier than summer bedding plants and are unlikely to suffer problems of drying out. Watering once a week or so will therefore be sufficient unless the weather is particularly mild and windy.

Evergreens are very useful in baskets hung in shady spots: box, elaeagnus, periwinkle, creeping fig and ferns are all successful in deep or semi-shade.

YOU WILL NEED

metal basket, 30 cm (12 in) in diameter
moss · plastic liner · 1 *Cotoneaster* 'Autumn Fire' · 1 greater periwinkle (*Vinca major* 'Variegata')
1 *Cotoneaster dammeri*

PERIOD OF INTEREST
October to April

LOCATION
In sun or shade, by a door or window, where evergreen foliage is lacking.

FOR CONTINUITY
Plant out the cotoneasters in the garden and replace with primroses (*Primula vulgaris*) for springtime colour.

1 Collect your ingredients. You will need moss and a plastic liner to help to hold water. Make some slits in the plastic for drainage. Water the plants well. Then line the basket with moss, green side outermost.

2 Add the plastic liner, fill with compost. Plant 'Autumn Fire' to the rear. Trail the periwinkle and other cotoneaster over and through the sides. Add more compost and water well. Hang from a strong bracket.

GOOD ADVICE

- Remove the chains, if possible, to make planting easier.
- Feed the plants after a few months.
- Use versatile evergreens as a foliage backdrop to year-round planting, for autumnal schemes and in shady spots.
- If evergreen plants in a hanging basket appear to be drying out, immerse in a bucket of water until bubbles stop rising. The rootball is now moist.
- Do not water if the roots are frozen, otherwise the leaves will desiccate.
- Aim to cover the base of the basket by encouraging plants to trail through the sides.

1 Cotoneaster *'Autumn Fire'*
2 *periwinkle*
3 Cotoneaster dammeri

Late-season Special

At a time when most garden plants have died back and lost their colour and leaves, you can still bring cheer to dull autumn and winter months with lustrous skimmias and wintergreen, with their glossy green leaves and red buds, which look marvellously eyecatching beside the colourful red leaves of the leucothoe. The spreading feathery juniper gives contrasting shape and texture to the whole arrangement.

Leaf colour and texture are part of the joy of late-season gardening, although evergreen plants will give colour and strength to a display all year round. Once winter is over, you can add spring and summer flowers to contrast with the glossy foliage.

YOU WILL NEED

a reconstituted-stone trough, 60 cm (24 in) x 23 cm (9 in) · 1 Spanish juniper (*Juniperus sabina* var. *tamariscifolia*)
1 *Leucothoe fontanesiana* 'Scarletta'
2 *Skimmia japonica* 'Rubella'
1 wintergreen (*Gaultheria procumbens*)

PERIOD OF INTEREST
October to April

LOCATION
In sun or shade, on a windowsill. The effect is improved if surrounding foliage contains echoing shades of colour.

FOR CONTINUITY
In a sunny site, underplant with waterlily tulips (*T.* 'Ancilla'), with their red-centred, white flowers. On a shady sill, underplant with delicate snake's head fritillary (*Fritillaria meleagris*).

1 First plan your arrangement with the plants still in their pots. Consider the shape of the spreading juniper and the situation of your trough in relation to shrubs below or climbers growing around your window.

2 Half fill the trough with potting compost. Infill with the largest plants – juniper, leucothoe and skimmia – then the wintergreen. Finally, add compost to just below the top edge of the trough. Firm gently and water in.

GOOD ADVICE

- Terracotta and glazed ceramic containers can crack in winter, so you may need to move them in before a hard frost. Some, of course, are frost-proof.
- Choose a group of plants with foliage that is varied in both colour and texture.
- Plan for one plant to trail over and break the line of the trough.
- With heavy stone or composite windowboxes, either plant in situ or fill a plastic liner separately then slot it in.

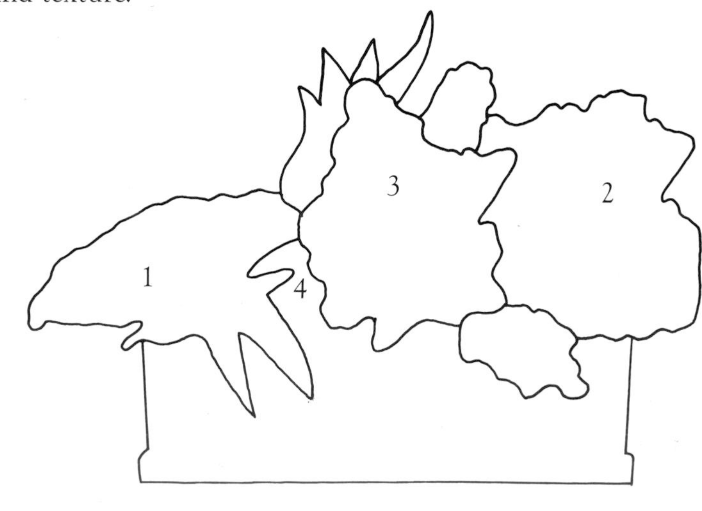

1 *juniper*
2 *leucothoe*
3 *skimmia*
4 *wintergreen*

Daisy Pot

The diversity amongst the daisy family is truly amazing and even if you had a hundred varieties you'd never get bored or want for colour, at least in summer and autumn when they're out in force. They're real charmers and well worth a pot to themselves.

You can plant up a daisy pot using easily available, off-the-shelf summer bedding plants, but I grew most of these from seed. It's more time consuming, but great fun and worth the effort, because you can be sure that no one else will have anything like it and you can tease your garden visitors with some of the more unusual varieties. By sowing in April and planting up in mid- to late June, your daisy pot will peak in late summer and autumn, when flower colour is becoming more scarce.

YOU WILL NEED

a terracotta pot, 55 cm (22 in) in diameter
9 coneflowers (*Rudbeckia hirta* 'Rustic Dwarfs') · 7 dwarf strawflowers (*Helichrysum bracteatum* 'Dwarf Hot Bikini')
7 *Oncosiphon grandiflorum*, syn. *Matricaria grandiflora* 'Gold Pompoms'
2 dwarf tickseeds (*Coreopsis tinctoria*)

PERIOD OF INTEREST
July to October

LOCATION
In full sun, surrounded by foliage as a foil to the bright colours.

FOR CONTINUITY
Aim for one big explosion of colour, then after the first frost concentrate on winter-interest plants.

1 Raise the annuals – coneflowers, strawflowers and oncosiphons – from seed in April. After 3–4 weeks, prick them out into seed trays, 24 plants per tray. In mid- to late June, almost fill the pot with potting compost.

2 Plant the tickseeds in the centre, then fill in around the back with coneflowers, carefully separated out from the seed tray. Put the strawflowers and oncosiphon around the edges and at the front. Water in well.

3 The plants will soon pick up as they make fresh root growth and become established. Then sit back and wait for the show!

GOOD ADVICE

- If you raise a seed tray of each of the annuals, you will have enough for fresh floral arrangements (coneflowers) and for drying (strawflowers and oncosiphons)
- Yellow *Coreopsis verticillata* would be just as good as red *C. tinctoria*.

1 *coneflowers*
2 *strawflowers*
3 *oncosiphons*
4 *tickseeds*

Autumn in the Air

You've got to admire michaelmas daisies. They respond to reduced daylight and falling temperatures and are genuine autumn-flowering plants. It is often November before they falter. Although not often associated with pots, the keen container gardener will want to make the most of these little gems. Another indispensable plant for summer and autumn is houttuynia. The heart-shaped, pink-and-yellow-splashed leaves make a terrific edging to a pot or trough and knit together in only a few weeks. *Fuchsia* 'Versicolor' sits well alongside with its arching shoots clothed with soft grey-green, variegated leaves and slim, pendant, purple, red-skirted blooms. Beneath this I slipped in a trailing knotweed whose leaves were taking on vivid autumn tints.

YOU WILL NEED

a plastic trough, 35 cm (14 in) x 55cm (22 in) · 2 dwarf michaelmas daisies (*Aster novi-belgii* 'Lady in Blue') · 2 *Houttuynia* 'Joseph's Coat' · 1 *Fuchsia* 'Versicolor' 1 knotweed (*Polygonum affine* 'Darjeeling Red')

PERIOD OF INTEREST
September to November

LOCATION
In sun or shade against a wall where a Virginia creeper or ornamental grape vine is colouring up.

FOR CONTINUITY
Add autumn crocus or colchicum bulbs along the front to strengthen the display.

Move the plant containers around until you have a satisfactory display. Then half fill the trough with compost. Insert the daisies at the back filling in with the other plants. Top up with compost and water. This arrangement looks better than a symmetrical one with the fuchsia in the centre.

GOOD ADVICE

- Pull out the trough plants in November and plant them in a border. If you wish to feature them again in pots, dig them up the following May.
- To avoid mildew, choose more resistant varieties like *Aster thomsonii* 'Nanus' and *A. amellus* 'Violet Queen' or spray every two weeks in summer with a fungicide.
- Don't forget to drill drainage holes in new plastic containers. Use a battery-powered drill for this.
- Why not use this trough as the centrepiece of an autumn display on a theme of gathering in the harvest. Set ornamental gourds, striped marrows and pumpkins around the base.

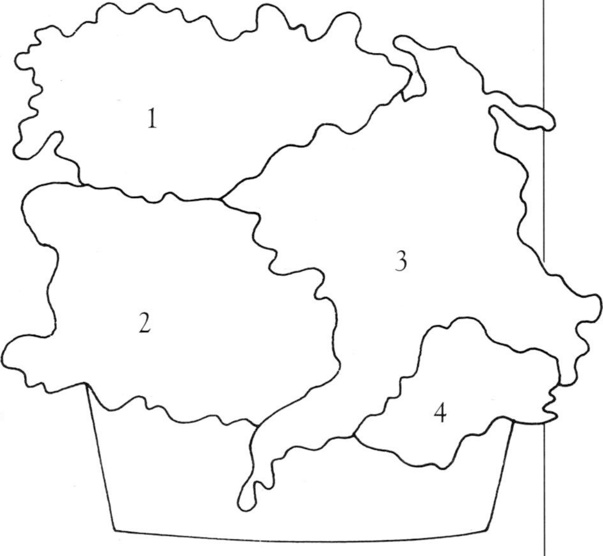

1 *michaelmas daisies*
2 *houttuynia*
3 *fuchsia*
4 *knotweed*

Two-tier Miniatures

Tiny treasures such as these violets and hardy cyclamen can easily be intimidated by outsized containers and overbearing neighbours, so I set out to make a mini landscape where they could be shown off in all their glory. It consisted of two clay pans, one perched on top of the other, with the lower one edged with a dwarf hedge of golden euonymus. When complete it looked as inviting as a wedding cake. To fill the gap between the euonymus stems and the pan side, and to give a decorated edge like piped icing, I slipped in houseleeks. These and the euonymus make a permanent frame that looks good throughout the year. Within the circle, plants can be swapped, so there is always something of interest.

YOU WILL NEED

2 terracotta pans 25 cm (10 in) in diameter and 46 cm (18 in) in diameter · 3 bedding violas · 3 *Cyclamen hederifolium* (syn. *C. neapolitanum*) · 1 *Sagina glabra* 'Aurea' · 4 pots of mini, gold-variegated euonymus (*E. japonicus* 'Microphyllus Aureovariegatus') · 18 houseleeks (*Sempervivum*)

PERIOD OF INTEREST
September to November

LOCATION
In sun or semi-shade, preferably raised up.

FOR CONTINUITY
Underplant the cyclamen with crocus and *Iris reticulata* bulbs to give colour in winter and spring.

1 Loosen the rootball of the euonymus, separating it into individual plants that can be spread evenly around the pan edge.

2 Assemble the plants. Fill the top pan with potting compost and plant it full of cyclamen and sagina. Then half fill the lower pan with compost. Plant the euonymus hedge towards the edge. Fill in with the violets, and finally insert the houseleek edging. Add more compost and water both pans well to settle the plants in.

GOOD ADVICE

- The best way to purchase hardy cyclamen is actively growing in pots. From dry corms they are slow and rarely as successful.
- Buy the variegated euonymus as rooted cuttings – several to a pot.
- Clip the mini euonymus hedge with scissors two or three times a year to keep it compact.

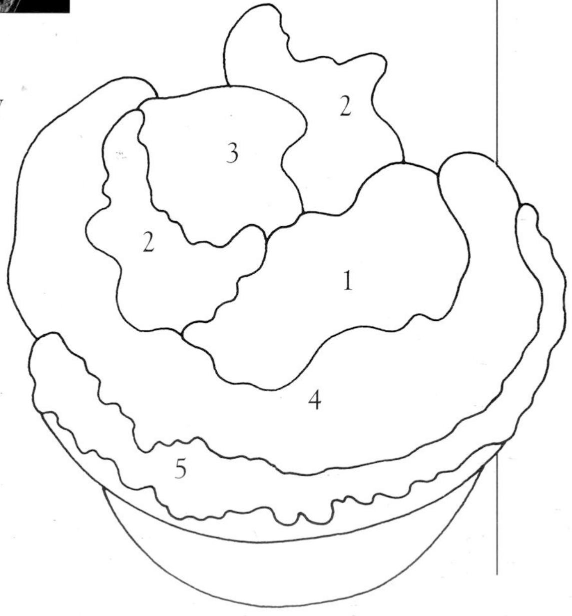

1 *violets*
2 *cyclamen*
3 *sagina*
4 *euonymus*
5 *houseleeks*

Seasonal Foliage

From the gold-tinged leaves of variegated holly to the luminous white of cyclamen and glossy berried evergreens, Christmas offers an opportunity for some delightful schemes.

You may never have thought of making a hanging basket specifically for the Christmas season, but among all the artificial tinsel and glitz it will seem refreshingly delicate and natural.

Roses can last for two weeks or longer if provided with water in phials. These can be wired and hidden among the greenery. Meanwhile the cyclamen will survive happily while temperatures are at freezing or just above.

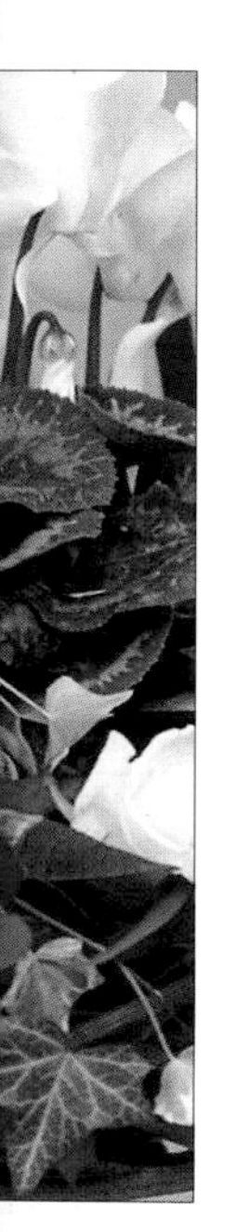

YOU WILL NEED

a woven-plastic twig basket, 23 cm (9 in) in diameter · plastic liner · 1 white cyclamen · 4 small ivies (*Hedera*) a woven-plastic twig wreath, 33 cm (13 in) in diameter · 8 plastic phials · 8 white roses · 1 stem of eucalyptus 1 Alexandrian laurel (*Danae racemosa*) 7 stems of snowberry (*Symphoricarpos*) 3 m (10 ft) of red ribbon, 2.5 cm (1 in) wide 1 m (3 ft) of red/gold, wire-edged ribbon, 6 cm (2 1/2 in) wide

PERIOD OF INTEREST
Mid-December to early January

LOCATION
In bright conditions, in a sheltered porch or cool hallway.

FOR CONTINUITY
Discard the flowers and foliage as they die; move the cyclamen to a cool windowsill.

1 Line the basket with slit plastic and fill with soil-less potting compost. Plant the cyclamen upright in the centre and trail the ivies over the sides. Top up with compost. Firm gently and then water carefully.

2 Cut the rose stems to 20 cm (8 in) and thread them through the wreath into water phials. Wire the phials. Thread the eucalyptus around the wreath, then the more delicate laurel and snowberries, inserting all these into the phials

GOOD ADVICE

- Small plastic phials of the type available from florists, and generally used for orchids and corsages, will give an arrangement such as this a lifespan of about two weeks.
- If a frost is forecast, bring the arrangement into a cool hallway for the night.
- Use wire-edged ribbon for impressive bows – it is easier to tie and far less floppy.
- If you wish for a more permanent Christmas basket without phials of water, simply combine the cyclamen and ivy in a container and hang it up on red ribbons.

1 *cyclamen*
2 *ivy*
3 *roses*
4 *eucalpytus*
5 *laurel*
6 *snowberries*

3 Tie the narrow ribbons on to the wreath to form a balanced arrangement when held from the top and tie the wreath to a cup-hook, or something similar, secured into a timber porch frame. Tie the wired ribbon into a large bow and pin it on to hide the top fastenings.

A festive red-and-white combination of roses and cyclamen wound about a rustic wreath with ivy, eucalyptus, laurel and snowberries.

Tree-filled Windowbox

Add to the seasonal festivities with a tree-filled windowbox full of Christmas charm. Dramatic red berries, glossy green leaves and symmetrical cypress trees, enhanced by the mellow warmth of a teak windowbox, make a Christmas arrangement that is both traditional and yet satisfyingly unfamiliar so that it will catch the attention of visitors and friends alike.

For too many of us, outside Christmas decorations are limited to fairy lights and a door wreath, when in fact there is a wealth of plants in the garden centre – with bright berries or leaves that have an interesting shape or texture – just waiting to be grouped in seasonal harmony. You could add extra decorations, such as silver or gold chains of stars, for Christmas week.

YOU WILL NEED

a wooden trough, 60 cm (24 in) x 23 cm (9 in) · ericaceous potting compost
2 *Skimmia japonica* subsp. *reevesiana*
4 small upright cypresses (*Chamaecyparis lawsoniana* 'Ellwoodii') · 3 ivies (*Hedera*)

PERIOD OF INTEREST
October to March

LOCATION
In sun or shade, beside a prominent doorway or on a windowsill.

FOR CONTINUITY
Underplant with 20–30 dwarf daffodils (*Narcissus* 'Tête-à-Tête') in autumn, or simply add pot-grown plants in spring.

1 Gather together your ingredients. Start by making holes in the plastic liner of the box, then add a small layer of grit or crocks to aid drainage. Half fill the box with ericaceous compost to maintain healthy growth.

2 Plant the cypresses towards the rear of the trough. Then firm in the skimmias, leaning their berries towards the front where possible. Infill at each end along the front with the ivies to soften the edges. Water in.

GOOD ADVICE

- Use berried plants other than holly in Christmas boxes to create an arrangement that is slightly unusual.
- Water weekly once the Christmas period is over and this windowbox is likely to last for a year or even longer.
- Despite being hardwood, teak windowboxes benefit from a treatment of wood preservative to keep them looking warm and mellow. A plastic liner will protect the interior.

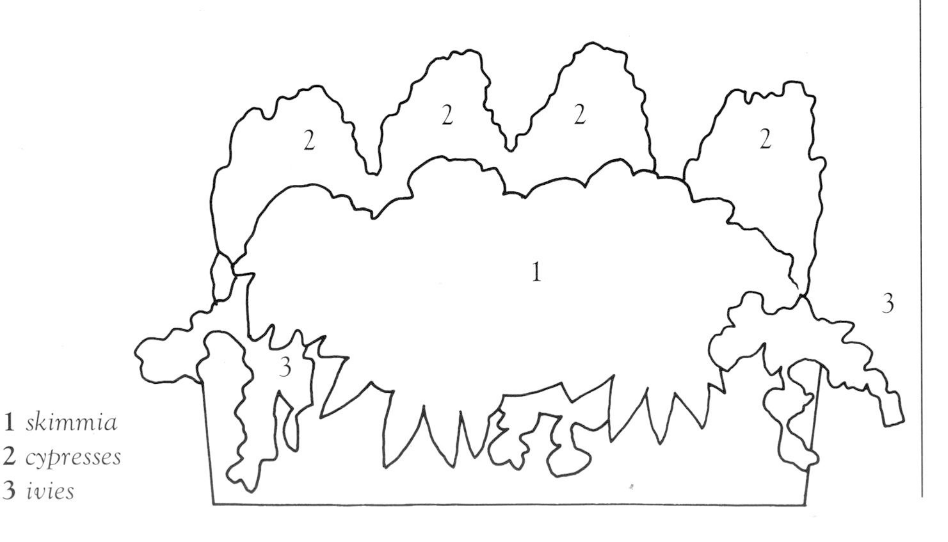

1 *skimmia*
2 *cypresses*
3 *ivies*

Christmas Trug

For a more unusual Christmas colour scheme, we decided to opt for gold and green, bringing inside plants that are often banished to the garden.

This arrangement could be used on a deep windowsill or even in a large fireplace. The plants could be moved to the garden after Christmas if you wish.

A few days before you plant up the trug, stain it in your chosen colour. We decided on spruce-green in order not to detract from the colour scheme and yet still be in keeping with a Christmas theme. Leave it to dry thoroughly for a few days, then line it with plastic. If you make drainage holes, you will need to water the trug outside.

YOU WILL NEED

a large wooden garden trug · ericaceous potting compost · 1 variegated osmanthus (*O. heterophyllus* 'Goshiki') · 1 spotted laurel (*Aucuba japonica*) · 1 dwarf thuja (*Thuja orientalis* 'Aurea Nana') · narrow gold ribbon · 1 large candle · 4 ivies (*Hedera*) · 1 *Euonymus japonicus* 'Aureus' 1 *Hebe elliptica* 'Variegata' · 5 dried quinces · 1 m (3 ft) gold cord · 1 gold cherub · 3 gold baubles · florist's wire

PERIOD OF INTEREST
December and January

LOCATION
In semi-shade, on a deep windowsill or in a cool fireplace.

FOR CONTINUITY
Plant out the osmanthus, thuja and spotted laurel and replace with yellow toning polyanthus.

1 Half fill the trug with ericaceous compost as the osmanthus, in particular, is acid loving. Insert the largest plants (the spotted laurel and the osmanthus), twisting the pots gently to remove the rootball intact.

2 Bearing in mind where the trug will eventually be sited, plant the hebe and thuja. Wind the narrow gold ribbon around the candle and choose a place for it where the flame cannot burn the trug handle.

3 Place the candle firmly in position so that it cannot fall over. Add the ivies and euonymus, then the quinces inside and out. Finally, wind the gold cord around the handle and wire on the cherub and baubles.

1 *osmanthus*
2 *spotted laurel*
3 *thuja*
4 *ivies*
5 *euonymus*
6 *hebe*
7 *quince*

Winter Warmer

Some plants look as though they have no business being outdoors in the dead of winter, let alone to be flowering their heads off. Tiny *Cyclamen coum* looks so delicate that it comes as quite a surprise to learn that it is in fact very tough. It also self-seeds very readily.

The best way to buy cyclamen is as young plants already in leaf and bud – not as dry corms that are reluctant to sprout and prone to rotting. I bought four *Cyclamen coum* from a local nursery in February, selecting them as much for their attractive leaf patterns as for their flowers.

It's important not to overpower such tiny treasures by putting them in enormous containers or with outsized companions, so here I selected some cheerful, dwarf, variegated evergreens.

YOU WILL NEED

a shallow, terracotta pan, 30 cm (12 in) x 20 cm (8 in) · 1 dwarf, variegated sedge (*Carex oshimensis* 'Evergold') · 1 variegated osmanthus (*O. heterophyllus* 'Goshiki') 1 dwarf, variegated euonymus 4 *Cyclamen coum*

PERIOD OF INTEREST
January to March

LOCATION
In sun or semi-shade, raised up where the pan can be seen from house, or on a balcony or in an unheated porch.

FOR CONTINUITY
Replace the cyclamen with pot-grown *Tulipa tarda* in early spring.

Choose a suitable pan of similar proportions to those shown here. There are several types on the market including glazed, bonsai pans. Fill it with potting compost to within 5–7.5 cm (2–3 in) of the rim. Plant the sedge so it will spill over the front. Position the osmanthus and euonymus at the back, and then fill in the gaps with the cyclamen still in their original pots. Top up the compost and water well.

GOOD ADVICE

- Stand the pan on four terracotta pot feet to improve its appearance. Such feet also prevent worms entering from below and blocking the drainage holes.
- To increase surface drainage, topdress the pan with small stone chippings.
- After flowering, you could remove the evergreens, plant the cyclamen properly and let them colonise the pan.

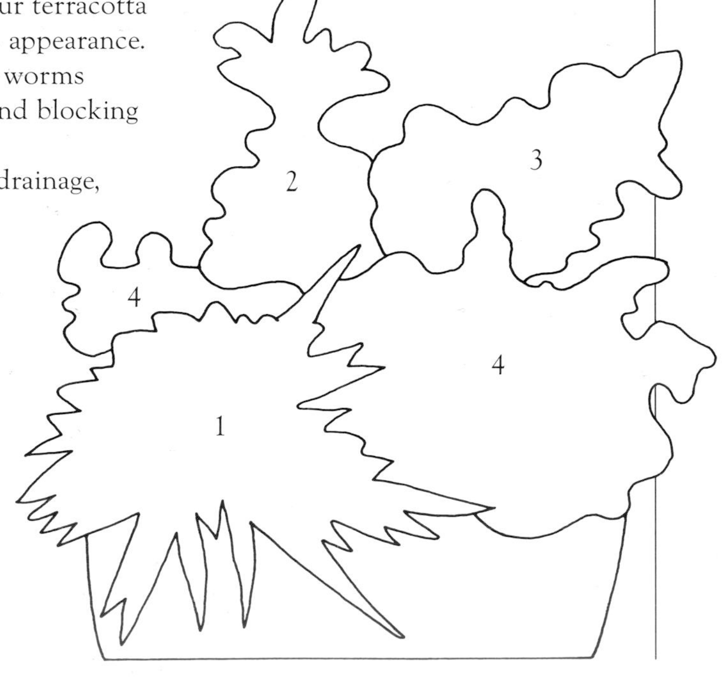

1 *sedge*
2 *osmanthus*
3 *euonymus*
4 *cyclamen*

Roses Revealed

Perhaps the most outstanding, winter-flowering herbaceous plant is the Lenten rose, which is far more likely to succeed in a container than the temperamental Christmas rose, *Helleborus niger*. The Lenten rose is hardy, flowers for weeks, and comes in a ravishing range of flower colours, some speckled or spotted. Its only drawback is that its nodding flowers tend to hang down, hiding their faces and the prominent stamens. To avoid this problem, plant it in a tall container like this Gothic font so you can greet it at eye level. To give additional support, station evergreens in front of your Lenten rose. I restricted this container to three colours – yellow, purple and pink – but nothing could outshine the ravishing Lenten rose.

YOU WILL NEED

a tall planter or urn, 38 cm (15 in) in diameter or sit a terracotta pot on another upturned pot or brick pier · ericaceous potting compost · 1–2 Lenten roses (*Helleborus orientalis*) · 1 female pernettya (in fruit) · 1 *Acorus gramineus* 'Ogon' 1 *Euonymus japonicus* 'Aureus' 12–16 crocus bulbs (*C. luteus* 'Dutch Yellow', syn. *C.* 'Yellow Giant') · moss

PERIOD OF INTEREST
January to early April

LOCATION
In sun or shade under a deciduous tree.

FOR CONTINUITY
Replace the yellow crocuses as they fade with primroses (*Primula vulgaris*) to maintain the yellow theme.

1 Assemble all the plants and, with the Lenten rose at the back, move the other containers around until the arrangement is visually satisfying.

2 Half fill the urn with ericaceous potting compost. Cut away all the old leaves from the Lenten rose before planting it at the back of the urn. (Fresh new leaves will quickly follow.) Then plant the pernettya, acorus, euonymus and crocus. Top up the compost and add a dressing of moss. Finally water everything well.

GOOD ADVICE

- The female pernettya will need a male plant nearby to produce berries for the following season. It also needs lime-free soil.

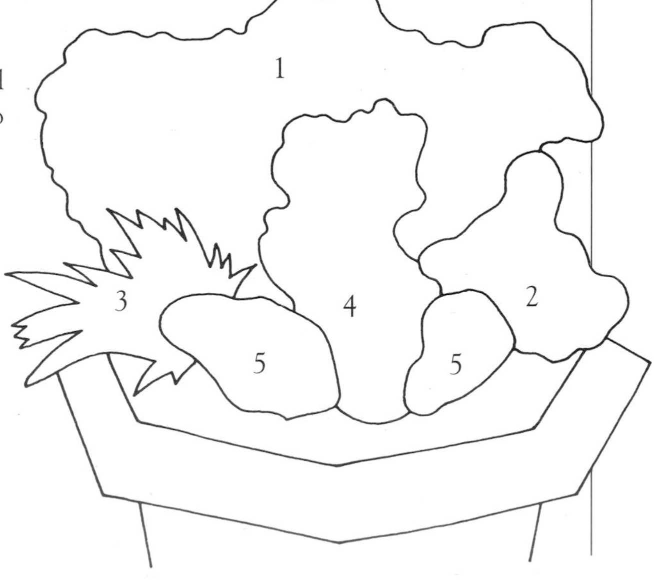

1 *Lenten rose*
2 *pernettya*
3 *acorus*
4 *euonymus*
5 *crocus*

Air Freshener

You might be surprised to learn how many winter-flowering shrubs, such as witch hazel, winter honeysuckles, viburnum and sarcococca (seen here), have a spicy, penetrating perfume, yet none has flowers bigger than a 10 pence piece. Conveniently, gardeners can harness these tantalising scents by planning a winter air freshener – a container packed with all the best that's on offer to flower all together in one glorious pot-pourri, or one containing flowers that bloom consecutively so that there is a different scent on offer every week or two. The basic framework of evergreens, with scented leaves, flowers or both, can be supplemented by a selection of dwarf scented bulbs, such as iris and hyacinths, and wallflowers.

YOU WILL NEED

a galvanised dolly tub, 40 cm (16 in) in diameter · polystyrene chunks or plastic bags · 1 *Choisya* 'Sundance'
1 sarcococca · 3 wallflowers (*Erysimum*)
1 trailing yellow conifer (*Chamaecyparis pisifera* 'Filifera Aureomarginata')
5 mixed pots of dwarf iris (*Iris reticulata*)

PERIOD OF INTEREST
February to early May

LOCATION
In a warm, sheltered spot, near a door or window, preferably raised up for effortless sniffing!

FOR CONTINUITY
Replace the iris with hyacinths or dwarf scented daffodils like *Narcissus* 'Baby Moon' to flower with the wallflowers.

1 For temporary displays, deep containers like half beer barrels or this dolly tub can be packed out to half their depth with polystyrene chunks or screwed up plastic bags to save on potting compost.

2 Gather the scented selection of plants together and check they are in good condition. Remove any dead or damaged shoots. Then add a good layer of potting compost to the tub, firming it gently to remove any air holes.

3 Plant the choisya, sarcococca, wallflowers, and trailing conifer, firming the soil around their roots. Carefully knock out the iris and slip them in amongst the evergreens. Top up the compost and water well.

GOOD ADVICE

- Keep an eye out for wilting leaves. Even in winter, frosts and cold winds can sap moisture from leaves and roots.
- Why not leave the yellow evergreen in place as the basis for a summer display.

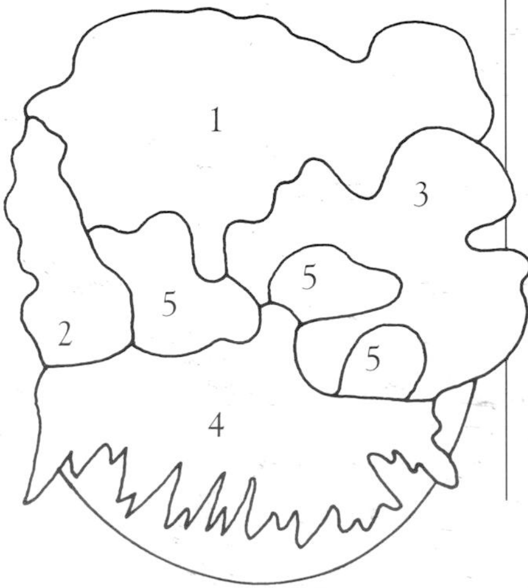

1 *choisya*
2 *sarcococca*
3 *wallflowers*
4 *conifer*
5 *iris*

The Festive Touch

That Christmas tree so eagerly anticipated, painstakingly selected and given pride of place in the lounge later becomes a waste-disposal problem. My 'Christmas tree' starts off as a snug potful of coloured evergreens planted up in November, dressed with home-grown garden decorations for the festive season and afterwards acts as the perfect host to show off potfuls of spring bulbs, polyanthus and pansies. Well, alright, it may not actually be in the lounge, but most homes have got a window or two where a special seasonal container, like this, can be laid on near enough to the house to allow armchair viewing.

YOU WILL NEED

a decorated, terracotta pot, 48 cm (19 in) in diameter · 1 juniper (*Juniperus* 'Grey Owl') · 1 specimen-sized variegated osmanthus (*O. heterophyllus* 'Goshiki') or variegated elaeagnus 1 *Chamaecyparis pisifera* 'Squarrosa Lombarts' · 1 tree heather (*Erica arborea* 'Albert's Gold') 1 *Skimmia japonica* 'Rubella'

OPTIONAL

honesty (*Lunaria*) · gladwin (*Iris foetidissima*) · old man's beard (*Clematis*)

PERIOD OF INTEREST
October to April

LOCATION
In sun or shade, on a balcony or porch or in a cool conservatory.

FOR CONTINUITY
Use dazzling, dwarf, red tulips like 'Red Riding Hood' to flower alongside the skimmia in spring.

1 The project got off to a flying start. A 'Grey Owl' juniper from a previous display was left *in situ* at the front of the pot, while as much potting compost as possible was changed without disturbing the juniper's rootball.

2 Position the taller evergreen at the back and bank down the heights saving the most prominent spot for the red-budded skimmia. Top up with more potting compost and water thoroughly to settle the plants in.

3 For a change of mood, add cut stems of dried honesty, gladwin and old man's beard. After dark you can bring the display alive with an uplighter skimming the outline of the pot.

GOOD ADVICE

- Avoid the gladwin if children might be tempted to eat its poisonous berries.
- A floodlight or outdoor fairy lights are also good accent lights.

1 *juniper*
2 *osmanthus*
3 *chamaecyparis*
4 *tree heather*
5 *skimmia*

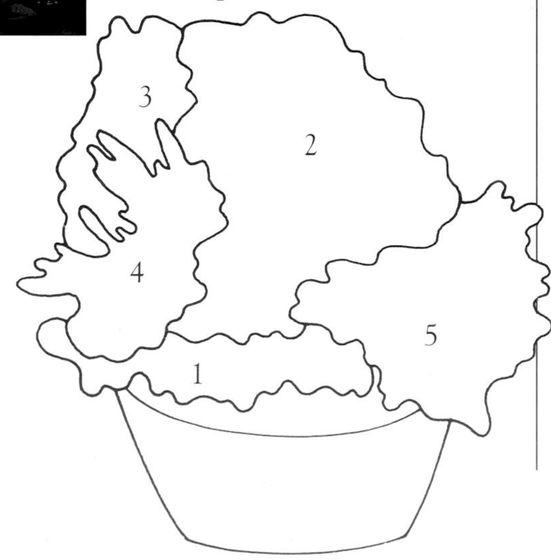

LOOKING AFTER YOUR CONTAINERS

When couples stop talking to each other it is often a sign that their relationship is breaking up, and it is the same with plants. Although at first this may seem strange, as you become more experienced you will begin to notice slight changes in, say, the colour and shine of the foliage, a decline in flower production or the failure to set fruit. Through close and regular inspections, your plants will tell you when they're unhappy and you will pick up the distress signals at an early stage, when they can be easily corrected. Don't always expect the obvious. A wilting plant is not necessarily dry at the roots. It could be waterlogged or possibly being eaten underground by vine weevil grubs.

When it comes to maintenance, hanging baskets are often a case apart: they require a regular watering routine, which may mean twice a day in sunny, windy weather. It is hard to overwater them. Conversely, a spacious wooden tub in a shady spot planted with, say, a Japanese maple may need only a weekly soaking. The more frequently you water, the more often you need to fertilise to replace nutrients leached out from the compost.

Few gardeners enjoy using chemicals to control pests and diseases, and early detection may enable you to hand pick caterpillars and squeeze greenfly. Spray as a last resort and, if possible, do so with an environment friendly product like soft soap.

A Simple Guide to Perfect Containers

Essential equipment

- Fold-up sheet for protecting paving as you plant up your containers.
- Hand fork for loosening compacted potting compost.
- Trowel for planting and grubbing out spent plants.
- Secateurs or scissors for pruning and deadheading plants.
- Watering can, preferably a long-necked, well-balanced one with a fine rose for watering seedlings.
- Trigger pump sprayer for cleaning leaves and a separate one for spraying pesticides. Write 'poison' on the one for chemicals and keep it out of reach of children.
- Gloves – some plants can irritate the skin.

Potting composts

There are two main types of potting compost: loam- or soil-based mixes such as John Innes; and soil-less potting composts, which traditionally are of peat, blended with sharp sand and a slow-release fertiliser. As environmental concerns for peat bogs deepen, coconut fibre (coir) and shredded bark alternatives are being introduced, which can give excellent results. Soil-based mixes are better for long-term planting and larger specimens. Soil-less composts are often suitable for sowing and potting; they are also cleaner and easier to handle than soil-based ones but do lose water and nutrients more rapidly. Sand, gravel, mica or vermiculite can be added to improve drainage.

Separate young plants by gently teasing apart the roots like this, the aim being to retain an equal amount of roots on each young plant.

Advance planning

- Soak new terracotta pots overnight by immersing them in a tank of water to prevent them drawing moisture out of the potting compost.
- Drill drainage holes in the base of new plastic containers, using a battery-powered drill. Plastic containers often have half-pierced holes marked out, and these just require a thin layer of plastic removing. Do not press too hard on the drill, or the plastic base may crack.
- Treat wicker baskets with two or three coats of clear preservative or two or three coats of waterproofing, yacht varnish.
- Set heavy stone pots and troughs into their final positions before you add any potting compost or plants. After planting, they may be too heavy to move.
- Set flat-bottomed containers on bricks or proprietary pot feet to raise them off the ground. This improves their appearance and prevents worms from entering and blocking the drainage holes.
- Use an ericaceous (lime-free) potting compost for acid-loving plants such as camellias and summer-flowering heathers.
- Before planting summer containers, mix water-retaining granules with the potting compost at the recommended rate to increase moisture-holding capacity.
- Wet potting compost that is dry and powdery until it is just damp

enough to stay in a ball when squeezed in the hand.

- For containers that will house the same plants for several years, fill up the bottom quarter of the container with drainage material such as crocks or gravel.

Planting up

- Start with clean containers, seed trays and pots; scrubbing them thoroughly in detergent will reduce infection from pests and diseases.
- Assemble your plants, working to a special theme, colour scheme or just a dazzling splash of colour! It is often a good idea to mix together plants of contrasting habit in the same pot so it fills out from top to bottom. Consider also whether you want an 'instant' display that looks good from day one, or one that will improve or peak with age.

With pot-grown bedding, make a hole for the plant in the compost. Carefully tuck in the plant and firm the compost using light pressure.

- Decide on which direction the pot will be seen from so there will be a definite front and back. There's no point in wasting trailers if no light gets to the back of the pot. Alternatively, as a centrepiece on a patio, it may be viewed from all sides so ensure interest all round.
- Avoid vigorous trailers in a container with attractive decoration on the sides, or it will be hidden.
- Make sure all your recipe plants are well watered before putting them into their display pot.
- Save on the amount of potting compost needed in large containers by filling the base with polystyrene chips from packaging material or old plastic bags scrunched up so as not to impede drainage.
- Lightly tap the rim of the pot on the workbench to loosen the rootball before taking out the plant and supporting it between your fingers.
- Position the plant with the deepest rootball in the pot first, having beforehand added sufficient potting compost in the base so that its level

If the rootballs of your pot-grown bedding plants are wrapped in tightly bound roots, gently tease out the roots to encourage new ones to form in the fresh potting compost.

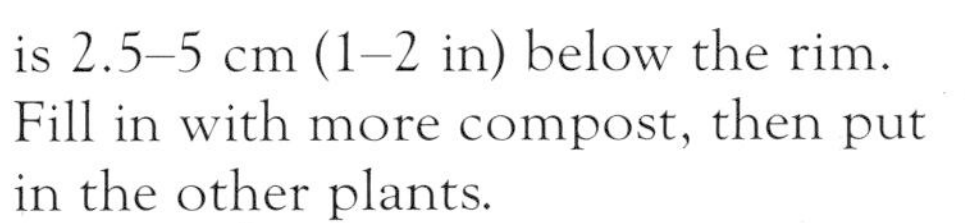

is 2.5–5 cm (1–2 in) below the rim. Fill in with more compost, then put in the other plants.

- Keep the soil off flowers and leaves as it is particularly hard to dislodge from hairy plants such as tobacco plants and petunias.
- Plants at the edge of a container display can be angled slightly to encourage them to drape over the sides of the container.
- After planting, carefully firm the potting compost around the plant roots. Then gently knock the container a few times on the ground to settle the compost.
- Add slow-release, push-in fertiliser pellets or sticks to the compost.
- Level the potting compost to within 2.5–5 cm (1–2 in) of the container rim so that the compost does not wash over the edge while the plants are watered.
- Sit back and enjoy your container once you've watered everything well!

Growing Tips

Bedding plants purchased as young plug or cell plants, potted up and grown on will cost less than half the price of more mature specimens.

Selecting a site

- Some container plants do well in shade, others demand sun, but many will grow well in both sun and shade.
- Avoid planting under trees such as limes and sycamores, because such trees attract aphids, which drop sugary sap on to the plants below. These then get covered in black sooty mould.
- You can override your garden's natural soil conditions by growing plants in containers. You can, for example, enjoy acid-loving plants such as rhododendrons in a garden with limy soil by growing them in a pot filled with ericaceous potting compost, which is acidic. Alpines that demand good drainage can even be cultivated in gardens with waterlogged clay by isolating them in a trough and using a free-draining potting compost.
- Place tender blooms such as camellias near the house walls, where they will be more protected – but avoid east-facing aspects.
- In exposed sites, choose compact plants. Top-heavy pots will be blown over. Also, use a soil-based potting compost, which is heavier than a soil-less one, and add pebbles or broken bricks in the base of the container for extra stability.
- In coastal regions, salt spray may scorch leaves so select salt-tolerant plants such as hydrangeas and fuchsias in such areas.

Good husbandry

- Plants now come in a range of cells, trays, packs and pot sizes. Those that are bought through mail-order firms need potting up as soon as possible on arrival.
- Buy fresh, healthy plants free from water stress, pests and disease. Roots should not be restricted or protruding from the base of the container, and the overall shape should be bushy and well balanced.
- Try growing your own plants from seed, selecting those to suit your site, lifestyle and colour scheme. Carefully follow the instructions on the seed packet, and sow in seed trays or individual pots depending on the plant's size.
- Water seedlings from above, with a fine rose, or from below to limit disturbance.
- Once their leaves are large enough to handle, prick out the seedlings into seed trays to give them more

space in which to develop.

- Young plants can be transferred directly from seed trays to your display pots or grown on in individual pots to produce larger, more bushy specimens that will often be coming into flower and so give instant impact.
- Old, used potting compost may harbour pests and diseases and be low on nutrients, so always use fresh compost each time you plant – or at least replace the top half with new.
- Trim off any broken shoots, dead flowerheads and seed pods and pinch back long, leggy shoots to encourage plants to bush out.
- Gradually harden off tender plants such as summer bedding and young plants to outside temperatures. Always bring them inside at night if frost is forecast.
- Planting tender plants outside is usually not recommended until the first week in June.
- Prune any dead, damaged or diseased parts on a plant to maintain the health of topiary, trained specimens, climbers and shrubs.
- Label herbaceous plants before they die back in autumn, to avoid accidental loss.
- In spring or autumn, repot perennials and replace the top layer of potting compost around established trees and shrubs to give them a boost.
- Pay particular attention to drainage in winter time, when containers can become waterlogged. Bulbs for example are prone to rotting.

Below *Take advantage of the flush of new growth in early summer to take cuttings and increase your stock of plants such as this yellow-leaved deadnettle.*

Above *Good-sized crowns of hostas like this can be sliced up into two or three separate pieces and grown on or planted straight into a display pot.*

Care and Maintenance

Watering

• Water containers using a can with a fine rose or a hose nozzle that gives a gentle sprinkling of water, not a powerful jet.

• A hose with a lance attachment (or a rigid extension and on/off switch) is invaluable to give you added reach up into baskets. Alternatively, a rise-and fall-attachment allows you to lower the basket for watering.

• Water in the morning or evening to reduce evaporation and avoid splashing water on to foliage in bright sunshine, or it may scorch.

• Watering cannot be done to a pre-set timetable. How often water is needed depends on the planting, the type of container and its siting.

• Water-retaining granules are particularly useful in hanging baskets as they increase the water-holding capacity of the compost so you need to water less frequently.

• Check each container daily in dry, sunny and windy conditions. Feel the compost with your fingers or weigh the pot by hand. Look for early signs of wilting.

• Watering is critical with hanging baskets, for once wind and hot sun have dried out the compost, they can be difficult to revive.

• Fill a dry container to the rim with water. Let it drain then repeat two or three times, or you'll wet only the surface even though water may appear through the bottom.

• Immerse a really dry container or hanging basket in a bucket or bath of water overnight. It may be the only way to get moisture back into the rootball.

• Containers, windowboxes and baskets will need less water in winter. Plants and soil will not absorb water when frozen solid, so if plants are wilting bring them into a shed or greenhouse to thaw, then water.

Feeding

• Feeding will keep your containers in tip-top condition, replacing nutrients used up by the plants or washed out through watering.

• Follow the manufacturer's instructions and don't overdose. Always apply a liquid fertiliser to moist compost. If dry, much of it will run off the surface and be lost down the sides.

• Solid or dry fertilisers can be mixed into the potting compost prior to planting, pushed into the surface, or applied as topdressings depending on their type. Some last for a whole growing season.

• Liquid feeds are concentrates, powders or granules sprayed through a special hose attachment or mixed up with water in a can. Apply every week or fortnightly.

• Foliar feeds act very quickly and correct trace element or mineral deficiencies.

• High-nitrogen feeds are good for foliage and potash ones are excellent for flowering, but a balanced fertiliser is best for mixed planting.

Maintenance tips

• Group your containers together to reduce maintenance time and to give mutual protection in windy or frosty weather.

• Regular pinching back of shoots will help plants such as fuchsias, pelargoniums, vigorous petunias and dwarf chrysanthemums (dendranthema) to bush out.

• Deadheading is essential for plants

Regular deadheading will prevent the formation of seed, which would slow down or stop the production of fresh flower buds.

such as nasturtiums, pansies, dahlias, cosmos and marigolds to maintain appearances and keep the blooms coming over a long period.

- Various hose attachments are available to aid feeding and watering as well as high- or low-level, automatic irrigation systems.
- Move plants to sheltered, shaded areas if you're going away on holiday and, if no one else can water, stand them on capillary matting kept damp from a reservoir bucket of water. Some plants such as monkey musk can be stood in trays of water.
- Use a wheelbarrow, sack truck or trolley to move heavy or awkward containers or try rolling them along. Do not strain your back.
- Containers such as imported terracotta that are not frost proof should be emptied and moved

Below *Settle in a newly planted pot by watering with the fine rose attachment on a can. If roots have been disturbed, stand the pot out of bright sun for a few days.*

Above *Protect the tender, young, red shoots of pieris with fleece at night if late frosts are forecast. Japanese maples will also appreciate this treatment.*

somewhere dry and frost free for the winter to prevent splintering or cracking. Alternatively insulate them well with straw, polystyrene, bubble wrap or sacking. This also protects a plant's rootball from freezing solid.

- Move tender plants indoors before the frosts begin in the autumn or root cuttings so you will have new plants for the following summer.
- Knock off heavy snowfall from evergreens such as conifers, or their shape may be ruined.

Pests and Diseases

To reduce significantly the incidence of pests and diseases, the golden rule is to keep your plants healthy through regular care and maintenance. Early detection can mean you merely need to squash pests in your fingers or pick off infected leaves. If you do resort to chemicals, be sure to select the appropriate one and follow the manufacturer's instructions. Plant-based extracts can provide some of the safest substances.

PESTS/DISEASE	SYMPTOMS	TREATMENTS
Aphids: blackfly, greenfly, whitefly	Ragged petals, distorted buds, stunted growth, sticky honeydew deposits, sooty moulds or fungal infections or deposits. Spread of viral disease during feeding.	Squash small infestations by hand, wash with soapy water or spray with an appropriate insecticide. Do not overfeed with nitrogen, because aphids prefer the soft new growth that it promotes.
Ants	Disturbed soil around roots; foraging on aphids and transferring them to fresh sites.	Not a serious problem; best left alone.
Slugs and snails	Holes in leaves, particularly hostas, bulbs, rhizomes, tubers and damaged shoots.	Go out at night or in wet weather and pick snails off; keep areas tidy to restrict daytime hiding places; or place inverted pots as traps, emptying each morning. For slugs, use pellets or shallow sunken traps with sweet, enticing solutions, which drown visitors. Raise containers off the ground and mulch with grit to slow down their movement.
Leaf miners	White or brown tunnels in leaves created by the larvae of moths or flies. Marigolds and chrysanthemums are particularly prone.	Pick off and destroy affected leaves.
Red spider mites	Discolouring or speckling on leaves, with fine, silky webbing; mites on the underside. More prevalent in hot, dry conditions.	Pick off affected leaves and spray with an appropriate insecticide.
Vine weevils	Adult beetles make notches in leaves while the wrinkled, white larvae with darker heads curl up in compost. Roots, rhizomes and corms are destroyed; plants become stunted or die; compost loosens and crumbles.	Tidy up to restrict hiding places of adults during the day, collect at night. Inspect pots in spring and late summer. Squash grubs or apply nematode, biological control.

Left *Search for snails hiding under leaves. Slugs prefer to creep into the soil or stay under a pot in the day.*

Right *As soon as tulips have gone over, pull them up. Put the petals in the dustbin to prevent an outbreak of tulip fire disease.*

PESTS/DISEASE	SYMPTOMS	TREATMENTS
Cuckoo spit	Frothy, white masses around stems protecting colonies of froghoppers. These sap-feeders distort young growth.	Squash insects by hand and hose off gently with water.
Grey mould (Botrytis)	Spots then fluffy mould and rot on flowers, leaves or stems. Common in wet conditions.	Increase ventilation, destroy affected areas, replacing badly diseased plants. Improve air circulation by removing lower leaves or spacing further apart. Spray with fungicide.
Powdery mildew	White, powdery fungus on leaves encouraged by dry conditions at roots.	Water regularly; pick off diseased sections; spray with fungicide.
Wilt	Plant wilting when potting compost is evenly watered; common to asters, zinnias, poppies and snapdragons.	Repot affected plant with fresh compost, or discard and replace with different species.
Rust	Raised brown, orange or yellow spots on leaves or stems; common on snapdragons and pelargoniums.	Remove and discard affected parts. Spray with fungicide.
Leaf spots	Diseased marks on leaves. Spots common to pansies; concentric rings to carnations; and blotches to columbines.	Pick off and destroy diseased sections. Spray with appropriate fungicide.
Virus	Varied symptoms affecting flowers, leaves or stems: from stunting and distortion to colour changes and mottling (mosaic virus). Few viruses are beneficial.	Untreatable so remove plant and destroy; do not use for cuttings. Control aphids to reduce transfer of infection.
Tulip fire disease	Brown, scorched leaf tips leading to rot; spots on leaves and flowers; black capsules on bulbs.	Gather and throw away or burn infected bulbs, petals and leaves.

The Ultimate Container Plant Directory

Star quality plants
Dwarf conifers
Dwarf daffodils (*Narcissus*)
Dwarf tulips (*Tulipa*)
Fatsia
Fuchsias
Geraniums (*Pelargonium*)
Hostas
Japanese maples (*Acer japonicum*, *A. palmatum*)
Lilies (*Lilium*)
Pansies (*Viola* x *wittrockiana*)
Tomatoes
'Yak' rhododendrons (*R. yakushimanum* and hybrids)

Scented flowers
Chocolate cosmos (*Cosmos atrosanguineus*)
Daphne odora 'Aureomarginata'
Deciduous azaleas (*Rhododendron*)
Heliotrope (*Heliotropium*)
Hyacinths (*Hyacinthus*)
Lily 'Star Gazer'
Petunias
Pinks (*Dianthus*)
Tobacco plants (*Nicotiana*)
Wallflowers (*Erysimum*)

Scented leaves
Chamomile (*Chamaemelum nobile*)
Curry plant (*Helichrysum italicum*)
Eau de Cologne mint (*Mentha* x *piperita* 'Citrata')
Lavenders (*Lavandula*)
Lemon balm (*Melissa officinalis*)
Lemon verbena (*Aloysia triphylla*)
Pineapple sage (*Salvia rutilans*)
Rosemary (*Rosmarinus*)
Scented-leaved geraniums (*Pelargonium*)
Thyme (*Thymus*)

Rapid growers for quick effect
Anthemis punctata subsp. *cupaniana*
Balcon ivy-leaved geraniums (*Pelargonium*)
Bidens
Cosmos
Helichrysum petiolare
Many hardy annuals
Mint (*Mentha*)
Sunflowers (*Helianthus*)
Surfinia petunias (*P.* 'Surfinia')
Trailing nasturtiums

Plants to perform a floral marathon
Begonia 'Non Stop'
Brachycomes
Busy lizzies (*Impatiens*)
Diascias
Felicias
Fuchsias
Lobelia
Marigolds (*Tagetes*)
Pansies (*Viola* x *wittrockiana*)
Scaevolas

Year-round interest
Acer palmatum 'Osakazuki'
Box (*Buxus*)
Choisyas
Fuchsia 'Thalia'
Hostas
Lenten rose (*Helleborus orientalis*)
Spurge (*Euphorbia characias* subsp. *wulfenii*)
Winter heathers (*Erica*)

Acid-loving plants
Azaleas
Callunas
Camellias
Gaultherias
Leucothöe
Pieris
Rhododendrons
Winter heathers (*Erica*)*
Witch hazel (*Hamamelis*)
* some species only

Shade-tolerant plants
Busy lizzies (*Impatiens*)
Cyclamen
Ferns
Fuchsias
Hostas
Ivies (*Hedera*)
Monkey musk (*Mimulus*)
Rhododendrons
Spotted laurel (*Aucuba japonica* 'Crotonifolia')
Tobacco plants (*Nicotiana*)

Drought-tolerant plants
Cacti and succulents
Festuca grasses
Houseleeks (*Sempervivum*)
Ivies (*Hedera*)
Lavenders (*Lavandula*)
Lewisias
Pines (*Pinus*)
Pinks (*Dianthus*)
Scented-leaved geraniums (*Pelargonium*)
Stonecrop (*Sedum*)
Thyme (*Thymus*)
Yuccas

Permanent container plants grown for their flowers
Angels' trumpets (*Brugmansia*, syn. *Datura*)
Camellias
Clematis
Fuchsias
Hebes
Hydrangeas
Pieris
Rhododendrons
Skimmia japonica 'Rubella'

Permanent container plants grown for foliage
Bamboos
Box (*Buxus*) topiary
Cordylines
Dwarf conifers
Euonymus
Fatsias
Ferns including tree ferns
Hostas
Japanese maple (*Acer japonicum*, *A. palmatum*)
Phormiums
Spotted laurel (*Aucuba japonica*)

Small trees
Aralias
Gleditsia triacanthos 'Rubylace'
Japanese maple (*Acer japonicum*, *A. palmatum*)
Sorbus x *hostii*
Sumachs (*Rhus*)
Weeping birch (*Betula pendula* 'Youngii')
Weeping cotoneasters
Weeping pussy willow (*Salix caprea* 'Kilmarnock')

Climbers*
Clematis
Cup-and-saucer vine (*Cobaea scandens*)
Honeysuckles (*Lonicera*)
Ivies (*Hedera*)
Morning glory (*Ipomoea purpurea*)
Summer jasmine (*Jasminum officinale*)
Sweet peas (*Lathyrus odoratus*)
Winter jasmine (*Jasminum nudiflorum*)
* some may eventually outgrow their pot

Trailing plants
Bidens
Brachycomes
Convolvulus sabatius
Diascias
Fuchsia 'Autumnale'
Helichrysum petiolare 'Limelight'
Lobelia
Petunia Surfinia 'Blue Vein'
Scaevolas
Verbena 'Pink Parfait'
V. 'Sissinghurst'

Exotic ornamental shrubs for a sunny patio*
Adam's needle (*Yucca filamentosa* 'Bright Edge')
Angels' trumpets (*Brugmansia*, syn. *Datura*)
Bottlebrush (*Callistemon*)
Cabbage palm (*Cordyline australis* 'Atropurpurea')
Camellia sasanqua
Coronilla valentina subsp. *glauca*
Dwarf pomegranate (*Punica granatum* var. *nana*)
False castor oil plant (*Fatsia japonica* 'Variegata')
Flowering maples (*Abutilon* 'Ashford Red', *A.* 'Kentish Belle')
Glory bush (*Tibouchina urvilleana*)
Karo (*Pittosporum crassifolium* 'Variegatum')
Lemon (*Citrus* x *meyeri* 'Meyer')
Lemon verbena (*Aloysia triphylla*)
Mimosa (*Acacia dealbata*)
Mimulus aurantiacus
New Zealand flax (*Phormium* 'Sundowner' and other striped varieties)
New Zealand tea tree (*Leptospermum scoparium* 'Red Damask')
Ozothamnus rosmarinifolius 'Silver Jubilee'
Parrot's bill (*Clianthus puniceus*)
Sweet bay (*Laurus nobilis*)
* overwinter in a frost-free greenhouse

Tender or half-hardy plants
Angels' trumpets (*Brugmansia*, syn. *Datura*)
Argyranthemums
Brachycomes
Cannas
Convolvulus sabatius
Dahlias
Diascias*
Felicias
Fuchsias
Geraniums (*Pelargonium*)
Helichrysum petiolare
Lobelia
Lotus
Osteospermums
Scaevolas
Trailing verbena
* some varieties are virtually hardy

Dwarf bulbs
Crocus
Cyclamen
Daffodils (*Narcissus*)*
Fritillaria
Glory of the snow (*Chionodoxa*)
Grape hyacinth (*Muscari*)
Iris reticulata
Scillas
Snowdrops (*Galanthus*)
Tulips (*Tulipa*)*
* some varieties only

Herbs
Basil (*Ocimum basilicum*)
Chives (*Allium schoenoprasum*)
Coriander (*Coriandrum*)
Curry plant (*Helichrysum italicum*)
Lemon balm (*Melissa officinalis*)
Marjoram (*Origanum*)
Mint (*Mentha*)
Parsley (*Petroselinum*)
Rosemary (*Rosmarinus*)
Sage (*Salvia officinalis*)
Sweet bay (*Laurus nobilis*)
Thyme (*Thymus*)

Ornamental salad vegetables to grow with herbs
Dwarf bush tomatoes
Ornamental cabbage
Ornamental kale
Peppers
Purple-podded, dwarf beans
Red-leaved lettuce
Ruby chard
Trailing tomatoes
Variegated nasturtiums
Yellow-podded, dwarf beans

House plants that can be stood outside in summer
Aspidistras
Begonias
Busy lizzies (*Impatiens*)
Cacti and other succulents
Christmas cactus (*Schlumbergera* x *buckleyi*)
Coleus
Jasmine (*Jasminum polyanthum*)
Spider plant (*Chlorophytum comosum*)
Tradescantia
Umbrella plant (*Cyperus alternifolius*)
Winter cherry (*Solanum capsicastrum*)

BERRY BARN
COTTAGE

FIXING BASKETS AND BOXES

How a hanging basket or windowbox is fixed in position will depend on its final location, design and even your choice of the plants themselves, so it's a good idea to consider the engineering side from the outset when planning a container display. The advantages of being able to position a windowbox where it can be enjoyed from outside and inside the house are obvious (it may also provide a degree of privacy from passersby), but if your windows open outwards, any windowbox may need to be fixed to brackets on the brickwork well below the sill.

Buildings with sliding sash windows often also have recessed sills, and here even the heaviest concrete planters can be housed, though you may need to consider their removal at a later date for house painting. Boxes on upper-storey windows must be securely tied into brickwork or the window frame, because they could cause serious injury were they to become dislodged.

Outsized hanging baskets may need hammer-in fixings or expanding bolts to make the brackets really secure, as will those arranged in tiers of three – like a wedding cake.

When fixing a hanging basket in position, always take into account how much space it will occupy once the display is mature, so that the basket does not obstruct doorways and other passages. Here now are some more fixing tips to get you off to a flying start.

How to Hang a Basket

Choosing a site

A spot that receives both sun and shade is best. Avoid north- and south-facing walls if possible, since these are often too shady or too hot. Also avoid a draughty corner.

A number of plants will thrive in shady areas. Take a look at our schemes on pages 98–101. Shade-loving plants for summer include begonias, bellflowers, busy lizzies, creeping figs, ferns, fuchsias, honeysuckles, hostas, jasmines, monkey musks, pansies, pick-a-back plants and tobacco plants.

In spring, bulbs such as lily-of-the-valley, daffodils, some tulips and snake's head fritillary will flower happily in semi-shade. Other spring-flowering plants that enjoy partially shaded sites include periwinkles, primroses, violets and azaleas.

Brackets

Wrought iron is the strongest and most suitable material. Choose a size to suit the diameter of your basket and to allow some space between your basket and the wall to prevent the basket from touching the wall and damaging the plants.

The bracket will have to cope with a heavy load, so secure it properly with screws and fixing plugs of the correct gauge and length. Check that they are fully tight from time to time throughout the year.

Basket size	Bracket size
25 cm (10 in)	23 cm (9 in)
30 cm (12 in)	23 cm (9 in)
35 cm (14 in)	23 cm (9 in)
40 cm (16 in)	30 cm (12 in)
45 cm (18 in)	30 cm (12 in)
50 cm (20 in)	35 cm (14 in)
60 cm (24 in)	40 cm (16 in)
75 cm (30 in)	45 cm (18 in)

Practical tips

Fixing up a bracket is very similar to putting up a shelf bracket. Baskets are very heavy when full – and especially after watering – and will need sturdy fixings. Most brackets come with the correct length of screw and wallplug. If you are fixing into render, a cavity wallplug, which splays out when inserted, is a good choice.

YOU WILL NEED

electric drill with correct size drill bit for screws
extension cable
screws – usually provided but otherwise 5 cm (2 in) long
wallplugs of the same length, standard or cavity depending on surface, i.e. brick or render
coloured crayon · measuring tape
straight or crosshead screwdriver

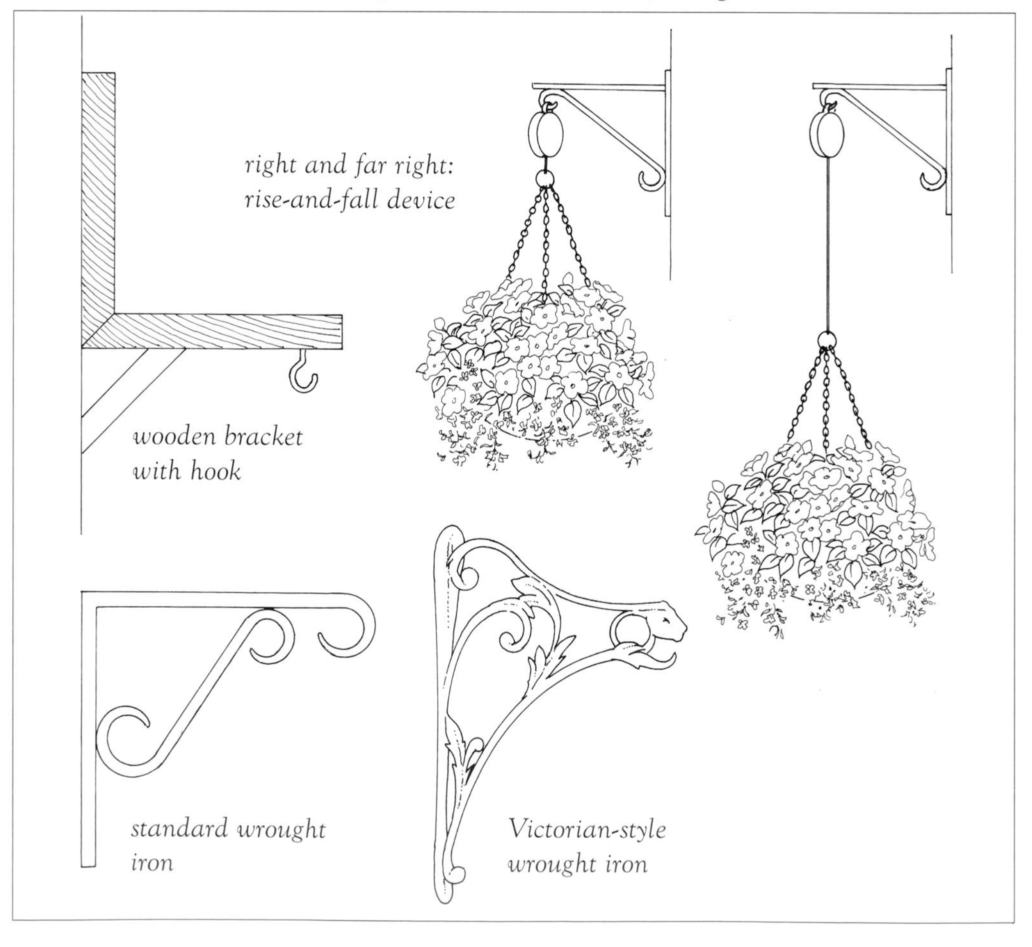

right and far right: rise-and-fall device

wooden bracket with hook

standard wrought iron

Victorian-style wrought iron

First ensure that the surface your bracket will rest on is relatively smooth. The top screw will be taking most of the strain, so check that this will be secured into a solid surface such as brick rather than into mortar.

Imagine the height that your basket will fall to once hung by the chains. Do not make it too high as you do not want to see only the underside of the basket. However, you also need to take into account whether it will be in the way of anyone walking by and if it is likely to be a danger. A heavy basket can cause an unpleasant accident if it is in an unexpected place.

It is most important to measure accurately as you do not wish to make more holes than necessary. The rule here is to measure twice and drill once! Mark the spots where you intend to drill with coloured crayon.

Then, holding the drill at right angles to the wall, maintain a light pressure when drilling. Drill in approximately 3 cm (1¼ in). (You can mark this measurement on your drill bit with tape or move the bit into the drill to the correct length.) Then insert the rawlplug and, holding the bracket in position, put in the screw. Leave tightening until both screws are in position.

When fixing into a wooden post or fence, carry out the same procedure but without wallplugs.

Fixing a windowbox

1 *Hang U-shaped brackets (bought to fit the windowbox) where there is no sill. Fix securely into the brickwork or render using wallplugs and screws of at least 5 cm (2 in) in length. Then screw the front of the windowbox to the brackets.*

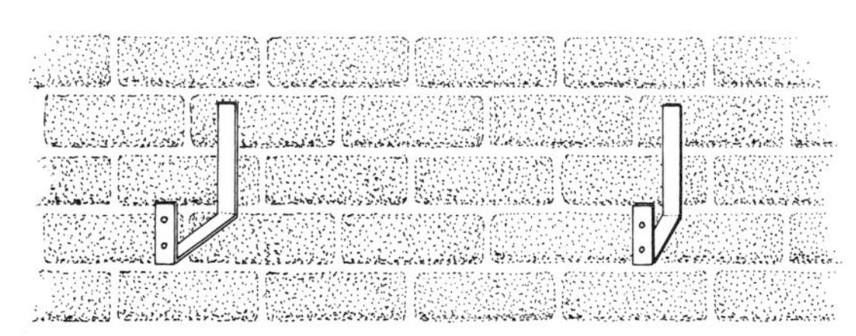

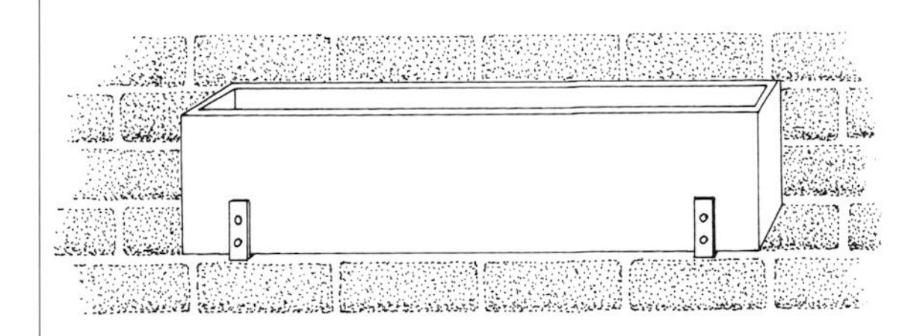

2 *If the box is to sit on a sill, fix L-shaped brackets to the base of the sill with the fronts projecting upwards. Sit the box on the brackets and screw them on to the front of the box.*

How to put up a Windowbox

Many windowboxes will sit securely on deep sills, but as they are so heavy it is still important to fasten them to the sill. If the box is wooden, you can use angle brackets at the front or sides, drilling and screwing these to secure the box. Otherwise, a length of decorative cast iron along the front of the sill makes an attractive safety option. Again mark first with a crayon and drill and screw the cast iron on, using wallplugs for a sturdy result.

Some sills slope and here you will need to place wedges beneath the box to ensure that it is level.

For windows that do not have sills, you can put up windowboxes underneath the window, secured by underneath brackets. Installation is much the same as when putting up shelves.

YOU WILL NEED

spirit level · measuring tape
crayon or pencil · 2 brackets for a box up to 90 cm (3 ft) long
3 brackets for a box up to 1.5 m (5 ft) long
4 screws and wallplugs per bracket
windowbox liner or crocks and plastic

Begin by marking out in pencil where the box is to be fixed. Then, spacing the brackets evenly along the base, mark where holes are to be drilled. Draw lines between the marks and check that they are in a straight line. Adjust where necessary. Drill holes using the correct size of bit for the screws. Place the brackets in position and insert the screws half way. When all screws are in, tighten up evenly.

Small plastic windowboxes, such as those used for children's boxes, can rest on two screws secured into a fence or wall.

Hanging Basket Liners

The first hanging baskets were made of twisted wire lined with sphagnum moss and filled with a peat-based or multi-purpose compost. For a large number of gardeners, this is still the preferred method as such baskets have a natural appearance, although today's wire baskets are usually plastic coated.

Many gardeners now include a polythene liner with slits inside the basket or they use a saucer to aid water retention. Various other devices also help to prevent waterlogging or drying out. For example, foam, recycled 'whalehide' paper, fabric and coconut-fibre liners are available with overlapping flaps and can be cut to fit any size of basket. These are usually available as overlapping petals, and planting through the slits of these liners enables the base of the basket to be covered.

Rigid cardboard liners are made of a compressed paper substance that is biodegradable and will not last more than one year. They hold water well but you cannot plant through the sides unless you punch holes in the liner.

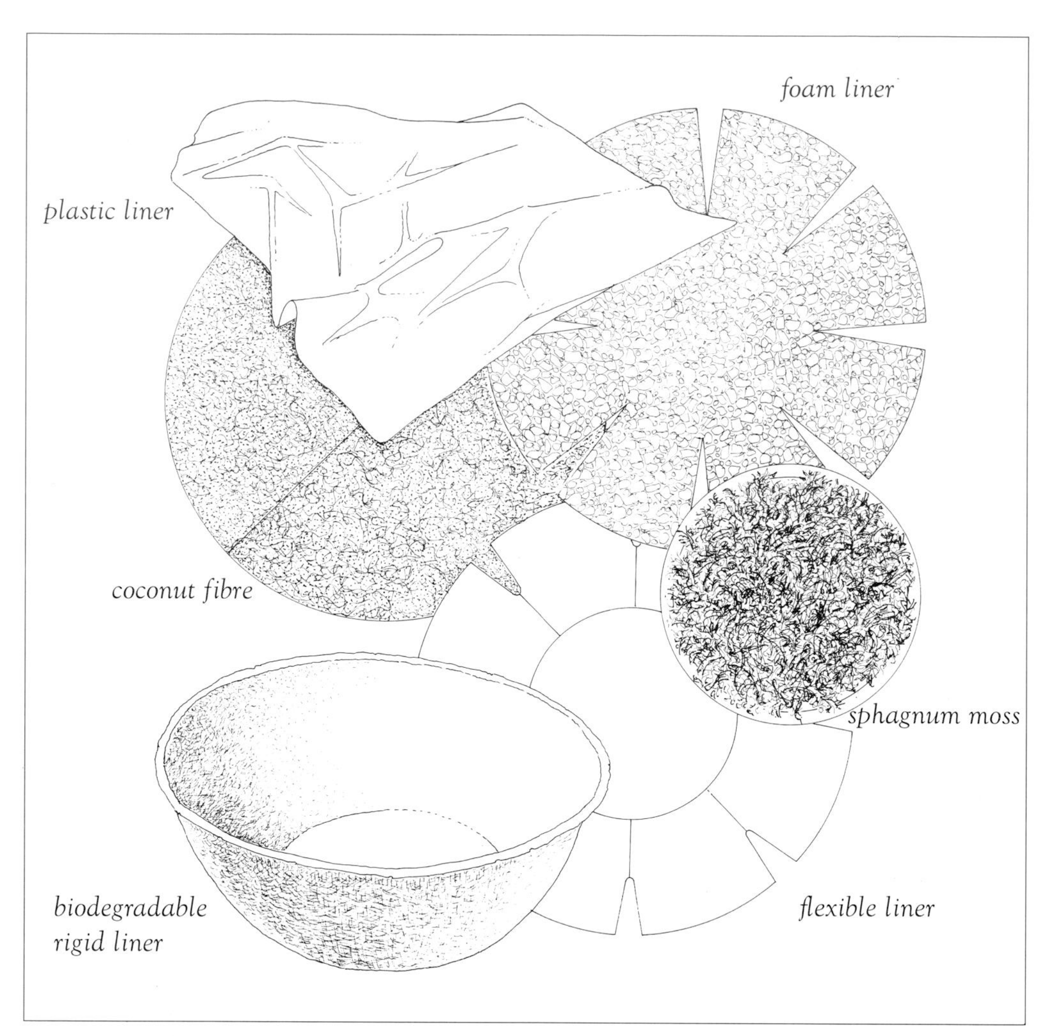

Solid-plastic hanging baskets are becoming more popular, particularly those with a false base or reservoir for conserving water. This overcomes one of the greatest problems encountered with hanging baskets – infrequent watering. It is not possible to plant through the sides, however, unless there are ready-made holes, so the use of trailing plants to disguise large areas of plastic is important.

Drainage Materials

Most windowboxes planned for outdoor use, including those made of terracotta, plastic, stone and composites, will not require a liner but should have drainage holes drilled into the base if there are none already. Cover these holes with a layer of crocks, which are broken-up terracotta pots. This prevents potting compost from being washed away. Large pebbles can also be used.

INDEX

AUTHORS' ACKNOWLEDGEMENTS

Claire Phoenix would like to thank the following people for their help:

Frankie Annis, Graham Wallis, Mr J. Wilson and Mr S. Curtis for horticultural help and advice. Also Mr and Mrs M. Fordham, Mr and Mrs J. Mustoe, Miss R. Amy, Mr and Mrs B. Barham, Mrs L. Allen and Mr P. Hiscocks, Mr and Mrs R. McLellan, Mr and Mrs R. Osborn, Mr and Mrs I. Fletcher, Mr and Mrs A. Knight, Mr and Mrs A. Tilbury, Mr and Mrs J. Wilson, Mr and Mrs Woore, Mrs J. Harper, Mr and Mrs Melville, Mr and Mrs C. Arnold, Mr P. Garnier, Mr and Mrs Huxley, Mrs H. Yeomans, Mrs J. Lewis, Mrs L. Fairholm, Mrs S. Ravenhill, Mrs W. Ellard, Dr E. Biggs, Mr and Mrs Mendelsohn, Saffron Walden nursery school and all the residents of Hampton, Middlesex for kindly allowing us to photograph their gardens.

Project editor Gill MacLellan for all her assistance with planting and photography, gardening editor Lesley Young and Maggie Aldred who art directed with care and precision. Also John Clarke at Garson Farm Garden Centre in Esher and Liz Gage, Caroline Owen, Sally Spicer and Dot and Caroline of Scotsdale Garden Centre, Great Shelford, Cambridge, for the supply of plants, containers and advice.

Special thanks to photographers Spike Powell and Glyn Barney for their enthusiasm and commitment. Photography by Spike Powell pp.14–17, 18 top left, 19 bottom, 20 bottom right, 23 top left, 24–5, 31 bottom right, 32 top right and bottom right, 33–4, 35 top left and bottom right, 36–7, 38 top left, 39–41, 47 top left, bottom left and bottom right, 62–5, 78, 80–83, 86–9, 92–7, 100–1, 104–21, 158–60, 162, 164–5, 172–7; Glyn Barney pp.18 top right and bottom left, 19 top, 20 top left and right, bottom left, 21–2, 23 top right, bottom left and right, 30, 31 top left and right, bottom left, 32 top and bottom left, 38 bottom left and right, 46, 47 top right, 84–5, 90–1, 98–9, 102–3, 161, 163, 198–9.
Additional photography by Eric Crichton pp.38 top right, 79; Stephen Crowdy p.18 bottom right; Clive Nichols p.35 top right and bottom left.

Chelsea Flower Show Exhibitors: Hermitage Horticultural Society, The Allotment Holders Association, The Harrow Show, Shipdham Horticultural Society, The Royal Hospital, Berkshire College of Agriculture, Brightling Flower Show Society, Horton Cum Studley Horticultural Society, Kingston Horticultural Society.

Specialist suppliers
Containers
Webbs, Unit 2, 15 Station Road, Knebworth, Stevenage, Herts SG3 6AP
Scotsdale Garden Centre, Cambridge Road, Great Shelford, Cambridge CB2 5JT
The Terracotta Shop, 8 Moorfield Road, Duxford, Cambridgeshire
Capital Garden Products, Gibbs Reed Barn, Pashley Road, Ticehurst, East Sussex TN5 7HE.

Plants
F. Annis, Abington, Cambridgeshire; Grace's Farm Shop, nr. Thaxted, Essex; Ickleton Trout Farm & Nursery, Ickleton, Cambridgeshire; Springwell Nurseries, Little Chesterford, Essex; Siskin Plants (alpines), Woodbridge, Suffolk; Squires Garden Centre, Twickenham, Middlesex; Putney Garden Centre, London SW15; Garson Farm Garden Centre, Esher, Surrey.

Sourcing pots, tubs and containers

Good places to buy pots, tubs and containers are garden centres, DIY outlets and garden shows or direct from the manufacturers. If you trace those made in the United Kingdom back to their source of origin you will get a valuable insight into how they are produced.

Containers used in the following recipes were provided by the manufacturer or supplier named:

Hyacinth parade (pp.66–7) ..... Bradstone Pots & Planters
Mellow yellow (pp.74–5) ...... Errington-Reay Ltd
Cool blue (pp.76–7) .... Bradstone Pots & Planters
Summer magic (pp.122–3) ..... Whichford Pottery
Summer pastels (pp.124–5) ..... Hampshire Garden Craft
Gracious herbaceous (pp.126–7) . Pembridge Terracotta
Dapper dahlias (pp.132–3) ..... C.H. Brannam Ltd
Cottage pot (pp.136–7)........ C.H. Brannam Ltd
A class act (pp.138–9) .... C.H. Brannam Ltd
Blazeaway begonias (pp.140–1) .. Bradstone Pots & Planters
Trail blazers (pp.148–9) .... Hampshire Garden Craft
Ice & lemon (pp.150–1) .... C.H. Brannam Ltd
Monkey business (pp.152–3) ... Errington-Reay Ltd
Roses revealed (pp.180–1) ...... Haddonstone Ltd

All other containers featured in this book are widely available from garden-product stockists, car boot sales, junk shops and auctions.

Photographs on the following pages were taken by Graham Strong: pp.1–2, 4, 7–13, 26–9, 42–5, 48–61, 66–77, 122–57, 166–71, 178–97.

Graham Strong would like to thank: Helen Griffin, Maggie Aldred and Joanna Chisholm for their ideas, enthusiasm and attention to detail; and Nicky Jones for helping with the care and maintenance section, and for growing and posing for photographs. Most of his recipes were grown using Arthur Bowers Multipurpose potting compost.

ADDRESSES

Contact for details of stockists/availability

Bradstone Pots & Planters,
Camas Building Materials,
Hulland Ward,
Ashbourne,
Derbyshire DE6 3ET
tel. 01335 372222

C.H. Brannam Ltd,
Roundswell Industrial Estate,
Barnstaple,
Devon EX31 3NJ
tel. 01271 43035

Errington-Reay Ltd,
Tyneside Pottery Works,
Bardon Mill,
Hexham,
Northumberland NE47 7HU
tel. 01434 344245

Haddonstone Ltd,
The Forge House,
East Haddon,
Northants NN6 8DB
tel. 01604 770711

Hampshire Garden Craft,
Rake Industries,
Rake, Petersfield,
Hants GU31 5DR
tel. 01730 895182

Pembridge Terracotta,
Pembridge,
Leominster,
Herefordshire HR6 9HB
tel. 01544 388696

Whichford Pottery,
Whichford,
Shipston-on-Stour,
Warwickshire CV36 5PG
tel. 01608 684416